D1765623

Food Flavourings

Food Flavourings

Edited by

P.R. ASHURST
Dr P.R. Ashurst and Associates
Hereford

Blackie

Glasgow and London

Published in the USA by
avi, an imprint of
Van Nostrand Reinhold
New York

Blackie and Son Ltd
Bishopbriggs, Glasgow G64 2NZ
and
7 Leicester Place, London WC2H 7BP

Published in the USA by
AVI, an imprint of
Van Nostrand Reinhold
115 Fifth Avenue
New York, New York 10003

Distributed in Canada by
Nelson Canada
120 Birchmount Road
Scarborough, Ontario M1K 5G4, Canada
16 15 14 13 12 11 10 9 8 7 6 5 4 3 2 1

© 1991 Blackie and Son Ltd
Softcover reprint of the hardcover 2nd edition 1991
First published 1991

British Library Cataloguing in Publication Data

Food flavourings.
1. Food. Flavourings
I. Ashurst, P.R.
641.338
ISBN-13: 978-1-4612-7838-2   e-ISBN-13: 978-1-4613-0499-9
DOI: 10.1007/ 978-1-4613-0499-9

Library of Congress Cataloging-in-Publication Data

Food flavourings / (edited by) P.R. Ashurst.
        p.   cm.
    Includes bibliographical references.

    1. Flavoring essences.   I. Ashurst, P.R.
TP450.F66   1990
664′.5—dc20   89-29401
                                                   CIP

Phototypesetting by Thomson Press (India) Limited, New Delhi

Preface

The flavour industry has become a vital element in the growth and success of food and beverage industries worldwide. The development of many new products is now directly related to the use of the appropriate flavouring which, among other benefits, has allowed the use of many novel raw materials. The phenomenal growth of specialised consumer products offering special tastes, nutritional benefits or 'convenience' almost always directly involves the use of a bespoke flavouring.

With recent growth in worldwide concern for environmental issues has come a corresponding concern for the use of 'natural' ingredients in foods. The flavour industry has been closely involved, by offering many of its products as natural alternatives, although the vexed issue of what 'natural' means has promoted discussion and debate in many quarters. The European Flavouring Directive has attempted to incorporate a definition. This is discussed further in chapter 1.

The work of the flavourist remains akin to that of the perfumer, despite inroads made by sophisticated analytical technology. For example, use of linked gas chromatography–mass spectrometry (GC–MS) instrumentation enables the skilled analyst to identify most components of a competitor's flavouring or the minor ingredients of a natural extract. Despite this, the industry remains a unique blend of art, science and technology in which the experience and knowledge of the flavourist is vital.

There are relatively few books about flavourings or the flavour industry, although many of the individual substances used are well documented. The industry remains highly secretive although mergers, technology and legislation have eroded many of the traditional secrets.

This book is intended to be a practical companion to the flavourist, the applications technologist and the technical sales person, and it will make a worthwhile contribution to the laboratories and libraries of all who are concerned with the manufacture or use of flavourings. It is intended as a source of basic information in a readable form, although it is not intended to provide formulation data. Individual formulations remain the lifeblood of the industry.

The book is in three main sections. The first section, chapter 1, is introductory and is concerned with the marketing of flavourings and legislative controls.

The second section, chapters 2–5, covers the main groupings of the raw materials of the industry. These are essential oils, natural extracts, fruit juices and perhaps most important of all, synthetic ingredients. The first three chapters of this section cover natural ingredients whilst the last covers both nature-identical and artificial components.

The third section, chapters 6–10, covers some of the main user industries, although the final chapter could in many respects stand alone. Chapters covering the user industries—beverages, confectionery, baking and the dairy industry—provide an outline of the technologies involved, where they are essential to an understanding of interaction with flavourings and their use. In some cases they cover the development of natural flavours within the products themselves.

The final chapter deals with process flavourings, which are playing an increasing role, and especially savoury flavourings. The chapter deals with their history, ingredients, methods of manufacture and application.

It is acknowledged from the outset that the book will have its shortcomings, if for no other reason than the sheer impossibility of covering all the aspects of a subject of this magnitude in one volume. Although I accept full responsibility for its faults, the book is the creation of its authors, all of whom are acknowledged experts in the practical application of their subject. I am indebted to them for their contributions and tolerance.

P.R.A.

Contributors

P.R. Ashurst Dr P.R. Ashurst and Associates 1 Stopford Close, Hereford HR1 1TW, U.K.

D.G. Ashwood Burtons Gold Medal Biscuits, Blackpool, Lancashire, U.K.

D. Bahri Givaudan Aromen GmbH D-4600 Dortmund, West Germany

E. Cowley 4 Ruskin Dene, Billericay, Essex CM12 0AN, U.K.

J. Knights PFW Ltd, PO Box 18, Greenford Middlesex UB6 7JH, U.K.

H. Kuentzel Givaudan Research Company, CH-8600 Dubendorf, Switzerland.

D.V. Lawrence Flavex Ltd, Goosesfoot Estate, Kingstone, Hereford HR2 9HY, U.K.

A.C. Matthews Net Consultancy, Cinderford, Gloucestershire U.K.

C.G. May H.G. Stringer and Co. Ltd, Tring, Hertfordshire, U.K.

D.A. Moyler Felton Worldwide, Bilton Road, Bletchley, Milton Keynes MK1 1HP, U.K.

S. and G. White The Edlong Company Ltd, 7 Anson Road, Martlesham Heath, Ipswich, Suffolk IP5 7RG, U.K.

J. Wright Bush Boake Allen Ltd, Blackhorse Lane, London E17 5QP, U.K.

Contents

4 Fruit Juices   87
P.R. ASHURST

5 Synthetic Ingredients of Food Flavourings   115
H. KUENTZEL and D. BAHRI

10 Process Flavourings

C.G. MAY

Appendices

References

Index

1 Introduction

E. COWLEY and J. KNIGHTS

1.1 General Introduction

In broad historical and traditional terms, the Flavour Industry has consisted of the blending houses who have created, manufactured and sold flavourings to the final user in end-products. In the context of the major forces shaping the present and the future of the industry, consideration must be given to:

(a) industry competition
(b) new entrants
(c) substitute products
(d) bargaining power of suppliers
(e) bargaining power of purchasers

Such a strategic analysis should identify the position of the industry within society and, in turn, the position of a company within the industry. This should reveal advantages and disadvantages for the business in respect of competition, no matter from which direction it will appear.

1.1.1 The U.S. Flavour Industry

Dorland and Rogers in their book *The Fragrance and Flavour Industry* [1] stated: 'Fifty years ago the essential oil and aromatic chemical industry of the United States consisted of about seventy firms, three quarters of which are no longer in existence. Over fifty of these companies were located in New York City, most of them in a small area of two or three blocks on the east side of Lower Manhattan, close to the docks.'

Of the seventy suppliers of essential oils and aroma chemicals listed in 1927, only the following were quoted as still doing business under the same name in 1978, and most can identify further changes in the decade 1978–1988.

Felton International Inc.	Compagnie Parento Inc.
Florasynth Inc.	Polak's Frutak Work Inc.
Fritzsche, Dodge & Olcott Inc.	Rhodia Inc.
Givandan Corporation	Roure, Bertrand Duport Inc.

D.W. Hutchinson & Co   Synfleur Scientific Laboratories
Hymer Aromatic & Import   A.M. Todd Company
J. Manheimer Inc.   Ungerer & Company
Neumann-Buslee & Wolfe Inc.   Van Dyk & Company Inc.
Norda Inc.   Albert Verley & Company

Other firms had undergone mergers or name changes, e.g. IFF, Lueders, Gentry; others have disappeared.

On the other hand, there was the appearance on the American scene of major companies such as Dragoco Inc, Firmenich Inc, Haarman & Reimer Corp. and others, none of which were apparent in 1927. Some very interesting individual company histories are reported by Dorland and Rogers [1] together with some potted biographies of the industry characters of the period.

1.1.2   The U.K. Flavour Industry

A similar picture is drawn in the United Kingdom and it is illustrated by comparing the members constituting the British Essence Manufacturers Association (BEMA) in the years 1917 and 1988.

Members in 1917

Clayton & Jowett Ltd   A. Boake-Roberts Ltd
W. Meadowcroft & Son Ltd   Stephenson & Howell Ltd
Duckworth & Co Ltd   John Stow
Burgoyne, Burbridges & Co Ltd   The Confectioners, Veg Colour &
W.J. Bush & Co Ltd   Fruit Essence Co
Manchester Chemical Co Ltd   C.W. Field Ltd
Barnett & Foster   John Cummins
                                    Bratley & Hinchcliffe

Members in 1988

Duckworth & Co Ltd   I.F.F.
Bush-Boake-Allen Ltd   Lucas Ingredients
Barnett & Foster Ltd   Fries & Fries
H.E. Daniel   Givandan
Dragoco Ltd   E.F. Langdale
Edlong & Co   Felton
Firmenich   P.P.F.
Cooke, Tweedale & Lindsay   P.F.W.
Dubois & Rowsell   Tate & Lyle
Eglington Yates & Co Ltd   Schweppes·
Florasynth   Pointings
F.D.O.   Zimmerman-Hobbs
Grindsted Products Ltd

Even during the period of 12 months 1988/1989 there have been significant changes in the membership of BEMA due to acquisitions, mergers and new entrants. Others have left the arena. The evolving story is similar to that occurring in the U.S. flavour industry.

What is not evident from the above is the rationalisation of products, product lines, customers and suppliers over the past 20 years. For example in the 1960s at least ten of the flavour companies had major product lines in compounded citrus juices with flavours and emulsions, but possibly only two have these products today. Self-sufficiency in raw materials has been rationalised for economic reasons but there are present day signs of a revival.

1.1.3 *The European Flavour Industry*

Development in the major continental Western European Countries parallels experience observed in the United States and the United Kingdom, i.e. the growth of the national industry, rationalisation, mergers and the invasion of the multi-nationals.

The membership list of SNIAA (Syndicat National des Industries Aromatiques Alimentaires) of France published in 1987 demonstrates some of the trends in the industry in that country.

Members in 1987

Aralco	Isnard
Aromex	Jaeger
Baube	Laurent
D. Blayn & Cie	Lautier Aromatiques
Camilli Albert & Lalque	Mane Fils
Colodor	Mero Rousselots
Daul	Metayer Aromatiques
Durban	Noirot
F.D.O.	P.C.A.S.
Felton World-Wide Fr	Quest International
Firmenich	Reynaud & Fils
Fontarome	Rhone-Poulenc C.H.
Gaget	Robertet S.A.
Gazan	Sevarome
Givandan France	Sima France
Granger Bouguet P.	Soco Fruits
Guedant	Vernier
Haarman & Reimer	

The presence of multi-nationals in the French market is significant, as some of these are not members of SNIAA. There are also good examples of traditional raw material suppliers searching for the higher added value of compounded products.

Moving to Eastern Europe, the growth of the national industry has not kept up with the pace observed in the West but the nuclei are in place for rapid growth once the demand is stimulated by the needs of the end-user industry for more volume of flavouring and greater variety of technological need. There is little doubt that once the problem of 'profit convertibility' is solved joint ventures in the East will be looked upon as a more desirable target by the international companies.

1.1.4 *The Far East Flavour Industry*

The most significant recent growth has been in the Japanese flavour industry and their expansion into the international markets as global players. However, parallel to this has been the intensification of effort of the multi-nationals into this area of high growth and comparative affluence linked to convertible currencies.

One of the challenges for the future of the flavour industry must be the Chinese market but selecting the right time frame to gain competitive advantage whilst not falling foul of restrictions is all important. Local production in this geographical area is evident and the emergence of new players in the overall flavour industry is also visible. This in some countries manifests itself through divisions of large industrial users.

1.1.5 *Classification of Flavour Companies*

From this brief survey we can see that the flavour industry (i.e. the industry composed of those companies which create, blend and sell flavourings) consists of members whose roots are varied and can be categorised as follows:

(a) Original flavour manufacturers with facilities for chemical synthesis, distillation/extraction and compounding. In some instances these companies have moved towards supply of general food additives as part of a horizontal integration rather than restricting themselves to flavourings alone.

(b) Original speciality chemical manufacturers who are moving closer to producing finished flavourings.

(c) Original end-users of flavourings who detected a 'business inside the business' who now manufacture, use and sell compounded flavourings.

(d) Original processors/importers of raw materials who have moved to blended products in search of higher added value.

(e) Original flavour manufacturers who have been acquired by larger industrial groups, e.g. chemical manufacturers. Presumably, such companies perceived advantages in the amalgamated businesses.

(f) Original food ingredient/additive supply companies in other functional areas (e.g. colourings, feed extracts) who perceived a more comprehensive and profitable service by adding flavourings to their range of products.

The supply position to the end-market industries has changed dramatically over the last thirty years both in the content of the players and also in their degrees of specialisation. The industry is best described as fragmented [2]. In adopting and applying Porter's theories to the flavour industry we conclude that it contains large tracts of mature business which is fragmented but containing competitive emerging niches, some declining niches and certain global opportunities. In most countries there are at least thirty flavour houses each striving for a national, if not international, role. As one result, the company which today finds itself in sole possession of a significant brief from a major customer is fortunate indeed.

The flavour industry is secretive about its own affairs and of necessity adopts a high code of confidentiality about its customers' projects. However the customer base is large, its needs are varied and generally the observed result is too many products, orders, customers and packaging requirements to obtain true business efficiency. Most companies have made efforts to rationalise these aspects to efficient levels but the overall problem remains.

1.1.6 Historical Development

Rosengarten [3] has presented a concise summary of the development in the use of spices. He claims the first authenticated records stated a use of spices in Egypt between the years 2600 to 2100 BC for embalming. In the first century AD spices found their way into the kitchen and up to 1200 AD Arab physicians were using them to formulate aromatic syrups with therapeutic properties.

In the dark ages it was the church and the monasteries who stimulated growth of formulated spices and then herbs. In the fifteenth century, both single spices and compounded formulations were being used to camouflage bad flavours and odours in food. The quantities and varieties were expanded over the next three centuries and covered the following examples: cassia, cinnamon, peppers, vanilla, clove, nutmeg, mace, saffron, oregano and some of the present day herbs.

During the years between the first and fourteenth centuries AD, formulations were devised for application to meat, bread, desserts and alcohol.

By the early nineteenth century an international business in spices and herbs had evolved in growing, drying, grinding and exporting. Seed and plant smuggling broke monopolies of growth areas. This development left a framework for a specific side of the flavour industry which still exists today even though the players/participants have changed. Extraction/distillation and increasing technological requirements by the manufacturing users has greatly expanded formulation and modes of presentation of products based on herbs and spices.

The expanding needs of users stimulated the movement of entrepreneurial chemists and pharmacists during the nineteenth century to further process other natural raw materials and also to investigate the synthesis of some of the major chemical ingredients of plants, fruits, herbs, and the extracts and oils derived from them. Hence the modern flavour industry was born in an environment of demand for its products due to a rapidly growing food and beverage manufacturing industry with increasing needs to meet technical and marketing innovations.

By the early twentieth century new parameters were becoming evident in the requirements of customers demanding research and development expertise of the flavour houses. Thus flavourings would have to be:

(a) stable to heat in aqueous media
(b) completely soluble in aqueous media
(c) uniformly dispersible in aqueous or oil phase or even throughout a colloidal matrix of a food
(d) capable of producing an agreed taste/aromatic profile
(e) unique—to provide the same uniqueness to the end-product (this could necessitate time spent in formulation or in raw material search)
(f) stable during prolonged storage

In these early years a new flavour profile could, for example, stimulate a confectionery company to market a new product line.

Over the past 30 years 'maturity' has become the descriptive word for the core of the flavour industry in general. Emerging niches are no longer the prerogative of the traditional industry and many new companies have been formed to exploit these niches and in some cases to challenge the traditional industry as well.

Since 1960, end-user industries have developed more and more new processes, ingredients, recipes, packaging and presentations. There has also been the sociological revolution for convenience, health and speed in every day life and in the kitchen. In the early 1960s, end-users were always anxious to look at new developments but from the early 1970s this attitude was changing dramatically. One major international food manufacturer is well remembered for his remark in 1973: 'if anyone is going to show me a raspberry flavour he will not be allowed inside the door—I have been submitted to 19 new variants this week.'

The top international flavour companies were probably all chasing the same major account and depending on the restrictions set, this could still mean up to 20 different submissions to the user company. It is not surprising that the user industry companies are now restricting the number of supply houses with which they will deal. Assessment of the capability of a flavour supplier to handle the parameters defined for the project necessitate not only provision of adequate resources but also continuity of communications. Under the same principles, the flavour company will be evaluating the opportunities for the chance of success, resources committed, time scale and potential reward.

1.2 Markets

The core business of the flavour industry is fragmented, i.e. no one dominates the field in a majority of markets. This is not surprising when a deeper look is taken at both the types of market and the geographical spread.

1.2.1 *Types of Market*

The user industry groupings are well known but variable and one which has proved a useful base for analysing opportunities is as follows:

Alcoholic drinks	Meat products
Animal foods	Margarine and fats
Baked goods	New foods
Beverages (not soft drinks)	Oral hygiene
Bread	Pet foods
Canned products	Pickles and sauces
Cereal products	Preserves
Chewing gum	Prepared foods
Confectionery	Soft drinks
Dairy products	Soups
Dehydrated foods	Starch products
Fish products	Snacks
Frozen foods	Tobacco
Fruit products	Vegetable products

By coupling this wide field of flavour users with the large number of flavour manufacturers, the wide variety of flavour profiles (including herbs, spices, juices, extracts, oils) and their parameters for technologically efficient application (standardisation, purity, ease of handling and measurement, long life, hygiene, legislative and labelling requirements), the reasons for fragmentation of the industry become apparent.

Another contributory factor to the fragmentation of the industry is the low capital cost for new entrants; very often a market advantage can be built round service and personal relationships.

A 1972 survey [4] of United Kingdom ingredient manufacturers as suppliers for the food industry showed the extent of the horizontal integration inside the industry (see Table 1.1). A similar pattern was apparent when the

Table 1.1   Food Manufacture Ingredients Survey, March 1972

Ingredient supplied	No. of manufacturers/ suppliers	No. also in FSSS[a]
Acidulants	16	6
Anti-oxidants	23	10
Cereal products	48	8
Colourings	38	10
Emulsifiers and stabilisers	84	9
Enzymes	23	4
Flavourings	82	
Fruit products	23	11

[a] FSSS flavour, spice, seasoning supplier.

ingredients supplier was extended to cover humectants, preservatives, sweeteners, thickeners, etc. A significant point from the analysis is that 11 flavour companies were not in the 1972 list, but today exist in specific niches of the industry.

In the face of this intense competition with its wide and varied market opportunities, where does a company find its opportunities for attaining business and increasing company profits? To gain this extra profit the company must first establish the presence of an advantage over possible competitors. The overall analysis will include the best status quo assessment of the following:

(a) the competition
(b) the user industry (special skills?)
(c) the specific customer (special relationship?)
(d) raw material supply (are you lean as a company?)
(e) pricing, profitability (is it relatively worth while?)

From the above assessment it should be possible to conclude whether a company has specific advantages or at least no disadvantage when winning the opportunity. Has the business the creativity to provide product differentiation, cost leadership, time scale and deadlines for submission? Can market preference be indicated or is there likely to be a little extra in the service compared with the competition? The size and growth potential of these industry markets has been well documented and further re-emphasis here will serve no purpose other than to state that savoury flavourings rather than the sweet profiles are growing faster at present.

1.2.2   Market Locations

In the initial search for opportunities, the other broad heading for assessment before formulating strategy must be geography. There is a great tendency today to look at the world market on a North to South basis. This results in identified regions:

(a) North and South America (including the Caribbean)
(b) Europe, Middle East, Africa
(c) Far East (including Australasia).

Various estimates have been made over the years and market surveys have been published for individual areas. In broad 1990 terms, the world-wide flavour compound market is valued at 3 billion U.S. dollars. Most of the multinational flavour companies of U.S. or European origin have been building up their operation in the Far East over the past ten years—an area of convertible money, open trading, quick decisions and fast growing GNP. However, these markets are highly competitive and likely to become more so and more expensive to serve in the future. The Far East and particularly Japan is an area of high innovation in the user industries. Other countries in the region are encouraging local production of flavourings.

Growth of volume and profit can be achieved by capitalising on a perceived advantage, e.g. service in development or delivery. All three regions contain countries which offer great opportunity with free trading and convertible currency but also great competition. At the other end of the scale they all contain countries where the competition is less but other problems can be daunting, such as:

(a) non-convertible currency payment
(b) extended letters of credit
(c) no majority shareholding in joint ventures
(d) profit repatriation restrictions
(e) fixed asset disposal problems

Perhaps the company which identifies quickest the changes in these areas leading to more liberal trading conditions may eventually make the largest gains.

1.2.3 New Entrants

The market was beginning to see fewer competitors in the flavour industry during the late 1950s and early 1960s, due to the spate of acquisitions and mergers which occurred. Since then there have been numerous new entries to the industry, particularly into identified niches and in some cases the publicity for the niches has been supplied by the industry itself.

The normally secretive nature of the flavour industry has been broken in the last 20 years due to the attendant publicity given to their collaborative efforts with governments in formulating flavour legislation. One consequent result has been that companies outside of the industry have seen the niches and identified their advantages over the traditional companies in exploiting them.

Similarly, the process of industry integration in the developed world and the general growth in the underdeveloped world has produced user industry companies whose purchases of flavours would be sufficient to constitute a medium sized flavour company turnover in its own right. This has led to an increasing number of user companies producing or investigating their own production of flavours. A logical development of this is apparent in a few instances, in that eventually sales of flavourings by them to other users will occur. Many manufacturers of flavouring ingredients and also raw material import/exporters are seeking greater added value, in consequence they too have moved into blending and compounding. As already indicated there has been horizontal integration by suppliers of other additives, e.g. colourings, into the manufacture and supply of flavourings. These developments occur because of perceived market advantages of service and value to the customer.

In summary, the markets for flavourings are highly fragmented and there are many opportunities which can be exploited by newcomers to the industry. The core business of blending and compounding is not 'fixed capital' intensive, and is not therefore a barrier to entry.

1.3   Products

The flavour industry today is market-driven and development is geared to measuring the needs of the industrial user markets.

1.3.1   Substitute Products

This is surely the major reason for the existence of the flavour industry. In the first instance, the industry seeks to achieve the out-performance of nature, for economic or technological reasons, to the benefit of the users. Once established, the competitive battle for perfection of flavour profile and, unfortunately minimum price, continues throughout the industry.

During the 1970s and 1980s, the wider application and greater sensitivity of analytical procedures for identifying trace ingredients both in nature itself and in competitive products, has led to a levelling upwards of 'quality of profile' throughout the industry. It is also arguable that the goal of the 'natural profile' is not the panacea it used to be. Sometimes the improvements achieved in this type of work are not significant enough to justify further time delay and increased cost in making the effort. Similarly with the increasing international choice of foods in the retail market, the consumer is beginning to appreciate the variability of nature itself.

1.3.2   Matching Competitors' Products

Opportunities do occur under this heading but it is always advantageous to pause and ask why. A few suggestions are listed below:

(a) Price reduction: If it is happening to your competitor it will happen to you sooner or later.
(b) Competitor is giving a bad service: It is useful to find out which part is bad, development, personal relations or deliveries. Make sure you can offer better.
(c) Growth of business and an additional supplier is needed.
(d) For specific geographical areas where you have advantage over present supplier, e.g. special trading relationships in another country.
(e) Formulation changes are required for legal or labelling reasons. Also possible due to raw material shortage.

1.3.3 Matching a Natural Product

There have been numerous examples in the past of natural flavouring materials being difficult to obtain; this shortage is frequently combined with very high price fluctuation. It is well to note that the flavour industry still uses predominantly natural rather than synthetic raw materials.

It must be remembered that all businesses are ultimately about investment and the level of return on it. The natural raw material side shows the move to large bulk growth and movement of fruits, herbs and spices. Difficulty is already being experienced in smaller volume crops as people have left the chain of grower–processor–importer–blender.

The need for matching or extending natural materials will be a growth aspect of the industry for the future. A few reasons are listed below:

(a) Wild/semi-wild crops are being harvested less particularly in the developed world.
(b) The cost of harvesting, even in the Third World, is proving too much for some processors even with cultivated crops.
(c) Ravages of war and self-inflicted ecological damage which is occurring throughout the world.
(d) Laboratory-made substitutes are improving and setting their own acceptable standards.
(e) Examples of demand exceeding supply of the natural crop are already known. These will grow in number.

1.4 Emerging Opportunities

1.4.1 General

Having stressed the fragmented state of the mature aspect of the business, a brief examination of examples of emerging opportunities and niches will be beneficial. Some of these are not new in principle but are capable of much wider exploitation than in the past. Others are based on the advanced borders of scientific knowledge and a deep assessment of investment, risk analysis, returns on capital is essential. A time frame for the development work against the availability of the market opportunity is advisable as competition may come from new entrants to the industry.

1.4.2 *Biotechnology*

The flavour industry is among those investigating the potential for biotechnology and the more open approach to 'natural' flavouring definition in certain geographic areas has given impetus to this topic. The subject is so full of potential that it is worthy of investigation whether or not the 'natural' label benefit is present. The legislative aspect is covered in Part 2 of this chapter and will reveal more about this specific point of labelling.

The areas where biotechnology can have the greatest impact on the flavour business are listed below:

(a) Production of natural flavouring ingredients, e.g. there has been a headlong dash for natural aliphatic acids with the consequent production of corresponding 'natural esters' from reaction with ethyl alcohol and amyl alcohols. This approach is capable of further extension but careful assessment of the parameters for viability need to be taken into consideration. Can the consumer expectation for informative labelling overtake any concensus of opinion agreed in manufacturing industry? Ultimately, Government rules and regulations must satisfy public expectation and opinion.

(b) Production of part or complete flavour, e.g. fermentation systems followed by physical methods of concentration (as in the production of butter flavour concentrates).

(c) Increased yields of essential oils, oleo resin and flavour components by the use of enzymes in natural material processing (e.g. addition of enzymes to onion processing).

(d) Plant cloning for maximum yields of secondary metabolites, e.g. the technology already exists for hybrid selection, which can be propagated at cellular level. There are well known examples in existing traditional herbs, spices and fruits where specific hybrids and growing conditions produce exceptional results over the norm. There are the beginnings of an advanced industry in this area. In fact the technology is in place and the field is open to those who can identify the specific areas to invest their money, management and commitment. This is an area of development where the advantages are lying with the flavour industry now for marketing research, management and promotion and for technology for growth, processing and formulation.

(e) Specific chemical steps to produce economies in the production of expensive flavour chemicals or to support natural claims, e.g. benzaldehyde production from other natural chemicals.

(f) Modified traditional fermentation systems to produce flavour enhanced 'soups' for direct use or further concentration, e.g. variation of the enzyme or substrate broths for maximisation of the result (e.g. enzyme modified cheese).

(g) Secondary metabolites from in vitro tissue culture of vegetable cells, i.e. the transfer of the plant growing process from the field to the laboratory.

The field of endeavour remains wide and varied but needs dedicated man hours plus investment for success. But who is best equipped to succeed? The market place too is wide and varied and it is considered opinion that the field is large enough for many players. In some instances the advantage will be with the specialised biotechnology company, in others with the plant hybridisation experts but there is also a wide field where the breadth of knowledge and experience in the flavour industry will remain paramount.

1.4.3 *Process Flavourings (see Chapter 10)*

These products cannot be described as emerging, as early patents in the field were assigned in 1960 (e.g. C.G. May, U.S. Pat. No 2,934,435). A savoury meat-like flavour was produced by reacting D,L-cysteine, furfuraldehyde and arginine. Numerous other patents appeared over ensuing years but the complexity and innumerable variations exploited in normal cooking processes indicates a large untapped store of riches for those with the commitment to work in this area.

A strong possibility for 'precursor type' flavourings exists where the reaction is made to occur within the food processing operation. The reverse can also apply where the food process is used to create the conditions of reaction. As an example, the feed of precursors to a snack extrusion process which gives appropriate conditions of time and temperature for reaction to occur and also builds in a carrier system, will develop the required flavour in situ.

1.4.4 *Technological Need*

The traditional parameters of temperature, time, base modification, uniform dispersability, and shelf-life are well known from the traditional user industries such as soft drinks, confectionery, biscuits, powdered desserts, etc. Newer processes such as UHT processing and extrusion with its high temperature/volume expansion, presents greater challenge of physical change and volatilisation. As another example, continuous automated meat processing is better served with water dispersible systems of flavourings than the traditional forms. New packaging concepts for final products can force changes in flavour formulation to meet the consumer demand for flavour profile, stability and shelf-life.

1.4.5 *Global Aspects of Flavourings*

These opportunities are manifested primarily through the raw materials rather than compounded flavours. The compounded flavour produced to supply food bulk as well as flavour, e.g. enzyme modified cheeses, tomato powder substitutes can also achieve universal status. It is worth noting that comparatively small raw material manufacturers have won a world-wide reputation because they supply a world-wide market and have a global position in a given raw material. On the other hand, there are some medium sized flavour companies who are not known outside their own country.

The scope for raw material specialisation is vast and no flavour company can be fully integrated for its total raw material requirement. However, the trend in recent years of moving out of raw materials must be reversed if a company wishes to remain as a global player. The solution lies in selecting the key raw materials to incorporate. Table 1.2 gives a broad outline of the category headings for raw materials.

Self-sufficiency in a raw material may be the catalyst for taking the first step but if internal usage is important it is also likely that there is a worthwhile external market for the same material. There are many examples of raw material profit margins being more attractive than those obtained on some formulated items. As in all commercial matters, the development depends on the advantage possessed but whilst compounded items are usually sold only once it is not unusual for a raw material to change hands two, three or even four times.

As companies vie with each other to be in the top four or five by the turn of the century, it is worth noting that the hypothesis 'only the large players will be able to afford the raw material base because of the capital investment required', may well be reversed.

Table 1.2 Sources of flavouring agents

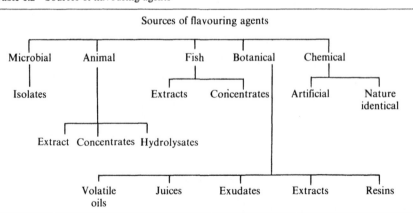

1.5 The Future

Even though the flavour industry is traditionally highly secretive, it has been the subject of a wide forum of investigation stimulated by legislation, consumer pressure and a compulsion on behalf of the industry itself to 'get it right' with these groups. This has led to disclosure of raw material lists, processes and in certain instances even formulations. This is in an industry where patent protection is not easily enforced except in chemical production processes.

The know-how is now breaking open with the result that secretive formulations of the past with high added value are becoming more like commodity items which can be purchased from many sources. This is the area of business in the industry where service and price are prime factors in gaining sales.

So what of the future? The market is not decreasing; it is still growing and each company has to analyse its own strengths and weaknesses in respect of the flavour industry, the user industries, the supplier, and legislative trends before formulating its strategy for the future. There is nothing difficult in this process although some best 'guestimates' may have to be used rather than absolute facts. In simple terms, to set out strategy, you are identifying a set of objectives for the company to reach and then putting in place the means or policies by which they can be achieved. These policies must be transcribed to functional plans and continuous monitoring instituted to see that the company goals are being addressed and that the functional roles reinforce each other. For example check that strategy is exploiting industry trends and opportunities against your position in it. Do the goals and policies deal with industry threats? Does the timing reflect environmental conditions? Checks must ensure that the resources are available, that communication channels operate, that the understanding of the objectives is clear and that the commitment and ability to succeed is in place.

The flavour industry is living through fast changing times and it is likely that the changes over the next 25 years will be greater than those of the last 25 years. How does the company take advantage of these changes? There are three basic strategic alternatives to consider:

(a) cost leadership
(b) differentiation
(c) focus

Cost leadership wins in mature, fragmented business; differentiation wins the emerging business whilst focus of effort can win in all circumstances. Whichever pathway is chosen, and it can be multiple choice applied to different segments of the business, the final analysis should indicate that the company is moving into a strongly advantageous position and that the objective can be achieved within an acceptable time frame. Many readers will be aware of the 'long term game plan' which eats resources until new management is instituted and axes it. The industry, itself, has been guilty of fostering competition but this result stresses the need to be better and to achieve individual company growth.

Sociological, environmental, legislative and labelling requirements will be factors to comply with but the skills to meet these needs and also those for the technological, toxicological and artistic performance are all possessed by the industry. There is no doubt it will respond to the challenges of the next decades; the interest lies in deciding which companies will perform best.

1.6 Legislation

1.6.1 *United States of America*

Until 1965 no country had a comprehensive flavour legislation although many countries controlled flavourings in the same way as other ingredients by having basic rules forbidding materials deleterious to health to be added to food. In 1965 the Food and Drugs Administration (FDA) of the United States decided to introduce legislation controlling all flavour additives by means of positive lists and requested the assistance of the flavour industry to achieve this. The U.S. flavour industry agreed to cooperate with the FDA by providing information on those flavouring ingredients that were currently in use. This was on condition that FEMA, the Flavor Extract Manufacturers' Association (the U.S. trade association) could set up an expert group to undertake the assessment of new flavouring substances. This was agreed to and the first 'GRAS' (generally regarded as safe) list was published [5] containing about 1100 natural and artificial ingredients. For this list no assessment was made since it just represented what was then in use. The stipulation by the FDA for future clearance was that an assessment of safety in use should be made by a group of independent experts qualified by their training and experience to make such a judgment. It is not necessary that every flavouring material which has been appropriately assessed is published in order to enable its use although the vast majority have been. Since 1965 a further eight supplements have been added to the FEMA lists [6], adding a further 600 mainly artificial substances. No distinction is made in the U.S. system between flavouring ingredients that are chemically identical to those occurring in food and those that do not, but natural ingredients are designated differently to non-naturals (artificials). One of the results of the GRAS system in a period when the number of known volatile materials in food has increased from about 500 to more than 5000, is that the number of GRAS flavouring substances has only increased from 500 to 1100. This has resulted in fewer flavouring materials being available to U.S. flavourists compared with those in the rest of the world.

However, because of the importance in the United States of being able to label flavourings as 'natural', a great deal of research effort has been put into trying to provide many of the familiar flavouring substances on the GRAS list from natural sources, such as by microbiological methods from other natural products or by isolation from obscure animal or vegetable source materials. It is not mandatory that the judgement required for GRAS status be made within the United States or by FEMA and it is interesting to speculate whether acceptance within Europe, judged by the European Scientific Committee for Food (SCF), will be acceptable to the FDA.

1.6.2 *The European Scene*

On 22 June 1988 the European Commission published an EC Directive on flavourings and source materials for their production [7] and a Council Decision on the establishment of an inventory of substances and source materials [8].

In the 1970s, West Germany and Italy had enacted flavour legislation based on a very limited positive list of artificial flavouring substances and negative lists of all other materials. This enabled new flavouring materials which were not artificial and posed no risk to the consumer to be used without publication or disclosure, making natural product flavour research worth undertaking, in contrast to the position in the United States.

The international flavour industry realised the significance of this position and set up an international body, the International Organisation of the Flavour Industry (IOFI) to try to influence future legislation. IOFI developed a code of practice for the industry [9] which made a distinction between flavouring ingredients based on the following ideas:

(a) Natural: Flavouring materials derived from animal or vegetable source materials, either raw or processed for human consumption, by physical or microbiological methods (such as solvent extraction, distillation or fermentation). There is no requirement that the source material must be able to be used as food.
(b) Nature Identical: Flavouring substances chemically identical to those which occur in materials normally considered as foods, herbs and spices.
(c) Artificial: Flavouring substances not yet shown to be chemically identical to those which occur in materials normally considered as foods, herbs and spices.

The Code of Practice also contains definitions for process flavourings, smoke flavourings and microbiological flavourings (still being developed), together with recommendations for manufacture, hygiene packaging and labelling. The code indicates in the labelling section that a natural compound flavouring may only contain flavouring ingredients classified as natural; the presence of a single nature identical flavouring ingredient renders the whole flavouring nature identical and the presence of a single artificial flavouring ingredient makes the whole flavour artificial. This ruling only applies to the flavouring ingredients themselves and not to added solvents, supports or other additives.

The reasoning behind the concept of nature identity is that the metabolism of food is in terms of the individual chemical species that are present in the food. Thus the metabolism of a single chemical substance is the same whether its source is a foodstuff or not. Therefore the risk to the health of the consumer is the same whatever the source, always providing that the substance is present in similar or lesser concentration in the flavoured food than in a normal food. This comparison with food has led to the concept of consumption ratio developed by Stofberg and Stoffelsma [10]. The consumption ratio (CR) is the ratio between the amount present in foodstuffs, processed for human consumption if appropriate, and the amount added to the food supply by the use of flavourings for the same population. Thus if the CR is high, say greater than 10, then the additional risk to health is negligible. Most nature identical flavouring substances are in this category. If, however, it is low, say, less than 1, then the substance should undergo toxicological assessment since it is adding significantly to the body load. The very common and widely used flavouring material, vanillin, is in this category with a CR of about 0.03. The CR together

Table 1.3   EC Maximum limits in food

Substance	ppm	Exceptions
Agaric acid	20	100 ppm in foods containing mushroom
Aloin	0.1	50 ppm in alcoholic beverages
beta-Asarone	0.1	1 ppm in alcoholic beverages and snack seasonings
Berberine	0.1	10 ppm in alcoholic beverages
Coumarin	2	10 ppm in confectionery and alcoholic beverages 50 ppm in chewing gum
Hydrocyanic acid	1	50 ppm in nougat, marzipan and subs. 1 mg/% vol. alcohol in alcoholic beverages 5 ppm in canned stone fruit
Hypericine	0.1	10 ppm in alcoholic beverages 1 ppm in confectionery
Pulegone	25	100 ppm in beverages 250 ppm in mint beverages 350 ppm in mint confectionery
Quassine	5	50 ppm in alcoholic beverages 10 ppm in confectionery pastilles
Safrole and isosafrole	1	2 ppm in alcoholic beverages <25% 5 ppm in alcoholic beverages >25% 15 ppm in foods containing mace and nutmeg
Santonine	0.1	1 ppm in alcoholic beverages <25%
Thujone	0.5	5 ppm in alcoholic beverages <25% 10 ppm in alcoholic beverages >25% 25 ppm in foods containing sage 35 ppm in bitters

with the quantity used has been incorporated into the methodology, developed by Rulis et al., for risk assessment of flavouring substances [11].

The most important aspect of the IOFI initiative is the recommended method of legislative control of flavouring materials, i.e. that artificials should be controlled by positive listing of those that may be used and that all other categories should be controlled by negative listing of those that may not be used because they pose some hazard to the health of the consumer (see for example, Table 1.3). This philosophy has been incorporated into the national legislation of Spain, Holland and Finland.

1.7 The European Flavour Directive

This framework directive sets out the categories of flavouring ingredients, other ingredients which may be added, limits for active ingredients and contaminants, labelling of flavourings sold for manufacturing purposes, but does not propose methods of control for flavouring ingredients of the labelling of foods containing flavourings. The main points of the directive are reviewed hereunder.

1.7.1 *Article 1*

'Flavourings' are defined as flavouring substances, flavouring preparations, process flavourings, smoke flavourings and mixtures thereof. It does not include solvents or supports or other additives necessary for the production or stability of flavourings as sold, but they may also be included (see Article 6).

'Flavouring substance' means a defined chemical substance with flavouring properties and is subdivided by its source and the method by which it is obtained as follows:

(a) 'Natural' obtained by appropriate physical, enzymatic or microbiological processes from material of vegetable or animal origin, either in the raw state or after processing for human consumption.
(b) 'Nature identical' obtained by chemical synthesis or isolated by chemical processes and which is chemically identical to a substance defined in (a) above.
(c) 'Artificial' obtained by chemical synthesis and not identical to a substance defined in (a) above.

'Flavouring preparation' means a product other than a flavouring substance, whether concentrated or not, with flavouring properties which is obtained by appropriate physical, enzymatic or microbiological processes from material of animal or vegetable origin, either in the raw state or after processing for human consumption. This definition includes extracts, essential oils, oleo resins, distillates and many other materials obtained directly from natural source materials.

'Process flavouring' means a product which is obtained by heating to a temperature not exceeding 180°C for a period not exceeding 15 min, a mixture of ingredients of which at least one contains amino nitrogen and another is a reducing sugar. It is not clear how a process flavouring that does not meet these criteria, such as a natural caramel flavouring, should be classified.

'Smoke flavouring' means a smoke extract used in traditional foodstuffs smoking processes.

As indicated above, each of these definitions has difficulties either in that they are imprecise or eliminate types of flavouring ingredients that are in common use such as rum ether, a non-natural flavouring preparation.

1.7.2 *Article 4*

Flavourings may not contain any substance in a toxicologically dangerous quantity. Maximum levels are provided for heavy metals and carcinogenic contaminants (Annex I). The maximum levels for pharmacologically active materials which occur in natural flavouring source materials are referred to and listed in Annex II.

1.7.3 *Article 5*

This deals with appropriate provisions concerning the groups of flavouring ingredients detailed in Article 1 but subdivides flavouring sources and nature identical flavouring substances into those related to foodstuffs, herbs and spices, and those related to non-foods. It does not make a similar distinction for natural flavouring substances. Presumably the appropriate provisions are concerned with the control of the flavouring ingredients. If this is so, the flavour industry, through its European trade association. Bureau de Liaison des Syndicats Europeens (CEE) des Produits Aromatiques, has suggested that the following provisions would be appropriate, achieving consumer protection whilst not stifling flavour research:

(a) Foods, herbs and spices as source materials, food derived natural flavouring substances, food identical flavouring substances, process and smoke flavourings should be controlled by negative lists; that is, lists of those materials which may not be used as source materials or flavouring ingredients because of their known toxic properties. This would be largely self-regulating because no flavour company would use any known toxic material in their flavourings at a level which would be deleterious to health.

(b) Non-food source materials, non-food natural and nature identical flavouring substances and artificial flavouring substances should be controlled by positive lists; that is, only those source materials and substances that are listed may be used in flavourings.

1.7.4 Article 6

This requires a positive list to be included in the Directive of non-flavouring substances that are authorised to be added to flavourings to dissolve, support or stabilize them. It also has to include processing aids that may be used. It is difficult to envisage why all food additives presently authorised for addition to food should not be used for flavourings and a list prepared of any additional substances that may be required solely for flavourings.

1.7.5 Article 9

This article deals with the labelling of flavourings sold for manufacturing purposes. In addition to the normal identification of manufacturer, packer or seller and the name of the flavouring, it requires a list, in descending order of quantity, of the categories of flavouring ingredients detailed in Article 1 and a second list, also in descending order, of the other substances present by name or EC number as required in Article 6. It also requires details of the maximum quantity of each component detailed in Annexes I and/or II or sufficient information to enable the purchaser to comply with the requirements of these annexes when the flavouring is used in a finished food. It is permissible to provide the above component information in the trade documents relating to the consignment rather than on the label itself provided that the statement 'intended for the manufacture of foodstuffs and not for retail' appears on the label. This is a sensible derogation since this part of the label information is of little use to the persons handling the container and would tend to detract from the possible risk and safety phrases on flavouring that are flammable, irritant or corrosive. The information concerned with Annexes I and II is necessary to enable the food manufacturer to comply with the Directive. It is somewhat doubtful what use can be made of any of the rest of the information unless it is in some way incorporated into the labelling of the finished food containing the flavouring. This seems very doubtful at present.

Article 9 also includes the rules for designating a flavouring as natural as follows:

(a) The word 'natural' or any word having substantially the same meaning may only be used if the flavouring component consists solely of flavouring preparations and/or natural flavouring substances.
(b) If the description of the flavouring contains a reference to a food or flavouring source, the word 'natural' or any word having substantially the same meaning, may only be used if the flavouring component has been isolated solely or almost solely from that source.

Paragraph (b) has many difficulties in interpretation. A previous European Commissioner publicly stated that 'almost solely' would, in his opinion, be satisfied by 90–95% from the named source, but this opinion has no legal backing. It is difficult to imagine how a flavouring which is totally natural as in (a) but does not conform to (b) should be labelled. Ideas for labelling such a raspberry flavouring as 'natural raspberry flavour flavouring' or 'natural flavouring—raspberry type' have been suggested but so far no decision has been taken by the Commission. It is difficult to visualise whether a designation such as 'natural barbecue flavouring' would meet the requirements of (b).

1.7.6   Article 13

This requires that Member States take the measures necessary to adopt the Directive within 18 months of its enactment; that is, by 22 December 1989, but very few states have managed to comply with this, mainly due to difficulties in interpretation. Flavourings complying with the Directive must be authorised by 22 June 1990 and those not complying be prohibited after 21 June 1991. Since flavourings which comply are already being sold the first of these requirements is largely meaningless, but the second demands adherence to the somewhat complicated labelling rules particularly in the matter of restricted substances.

   The other articles are relatively unimportant from an industry point of view being concerned mainly with the legal requirements of implementation and modification.

1.7.7   Annexes

1.7.7.1   Annex I.   There is presently only one substance in this annex, namely 3,4-benzopyrene, a well-documented carcinogen. The maximum limit of this substance that may be contributed to food by the use of flavourings (mainly smoke flavourings) is 0.03 parts per billion (ppb). This limit is unrealistic when it is compared with levels of benzopyrene occurring in smoked, grilled or fried foods (1–3 ppm) and the level of detectability (about 0.03 ppm). It is suggested that benzanthracene should be added to this annex since the procedures adopted to remove benzopyrene are not equally effective in removing benzanthracene due to its higher steam volatility. It is also suggested that the potentially carcinogenic substances (such as substituted quinoxalines) which are produced in process flavourings using creatinine containing materials, might be added to the list but this will prove impossible until adequate methods of analysis are developed.

1.7.7.2   Annex II.   This contains a list of twelve substances with adverse pharmacological properties that are contained in natural source materials and thus in flavouring preparations. These substances may not be added to food or flavourings as such and are limited in concentration in food due to the use of flavouring preparations. They are listed in Table 1.3.

1.8 Council Decision

This decision is concerned with the acquision of data which the Commission states that it requires in order to assess all questions relating to flavourings and source materials for their production and action to be taken as a result of the assessment. In order to acquire the data the Commission requires that an inventory be established of the following:

Flavouring sources composed of foodstuffs, herbs and spices
Flavouring sources composed of animal or vegetable non-foods
Natural flavouring substances
Food nature identical flavouring substances
Non-food nature identical substances
Artificial flavouring substances
Source materials used in the production of smoke and process flavourings
and the reaction conditions of their preparation

The information is being collated by Bureau de Liaison to be submitted by the Decision deadline of 22 June 1990. The size of the sections are approximately:

Flavouring sources   10000 entries
Natural and nature identical substances   5000 entries
Artificial flavouring substances   350 entries

There still remains the aspect of confidentiality of the information that is to be submitted. It is important that a mechanism be agreed with the Commission to allow data to be provided with the guarantee that it will not be divulged or published within an agreed time frame, say 5–10 years. Until this is agreed it is certain that a number of the flavour specialities discovered and used by individual flavour companies will not be submitted.

Whether the timetable for the implementation of the Directive and Decision will be adhered to is difficult to forecast. There is no doubt, however, that when it becomes law it will have a profound effect, not only on the flavour industry in Europe, but also on the rest of the world.

2 Essential oils

J. WRIGHT

2.1 Introduction

Most foods derive their characteristic flavour from chemicals which are present at levels ranging from parts per billion to parts per million. On the broad canvas of nature, some species evolved with far higher levels of flavour chemicals. Dried clove buds, for example, contain 12% eugenol. Such herbs and spices have been used from very early times to flavour other foods. With the discovery of distillation, it became possible to separate the flavour chemical mixture from the botanical material, and essential oils were born.

Essential oils and material extracted by solvent from herbs and spices found ready use in the rapid evolution of the soft drinks and confectionery industries. They formed the backbone of the raw materials used by the early flavour industry. Even today, essential oils form a significant part of the flavourists' repertoire and training normally covers the subject in great detail.

2.2 The Production of Essential Oils

2.2.1 Steam Distillation

Essential oils are produced by a variety of methods. Steam distillation is the most widely used (Figure 2.1). The steam is normally generated in a separate boiler and then blown through the botanical material in a still. The basic principle behind the distillation of two heterogeneous liquids, such as water and an essential oil, is that each exerts its own vapour pressure as if the other component were absent. When the combined vapour pressures equal the surrounding pressure, the mixture will boil. Essential oil components, with boiling points normally ranging up to 300°C will boil at a temperature close to the boiling point of water. The steam and essential oil are condensed and separated. Essential oils produced in this way are frequently different from the original oil in the botanical in a number of respects. Chemicals which are not volatile in steam, for example 2-phenyl ethanol in rose oil, are mainly left behind in the still. Many of these non-volatile components are responsible for taste rather than odour effects. Some very volatile chemicals may be lost in the

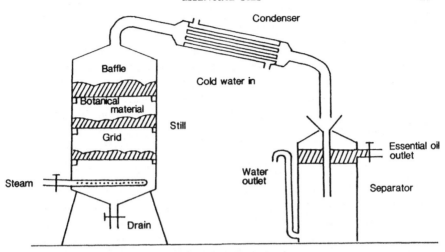

Figure 2.1   Diagram of typical steam distillation plant.

distillation, and the process itself may induce chemical changes such as oxidation or hydrolysis.

2.2.2   *Water Distillation*

Water distillation is similar in many respects to steam distillation except that the botanical material is in contact with the boiling water. 'Still odour' is a frequent problem with this type of distillation particularly if the still is heated directly. The burnt character of 'still odour' will gradually reduce on storage of the oils.

Steam and water distillation combines some features of both methods. The botanical material is separated from boiling water in the lower part of the still by a grid. If the still is heated carefully this method reduces the danger of 'still odour'.

Hydrodiffusion is a variation on the normal steam distillation process where the steam enters from the top of the vessel and the oil and water mixture is condensed from the bottom. This method reduces distillation time and is particularly suitable for distilling seeds.

Water distillation is the method of choice for fine powders and flowers. Steam fails to distil these materials effectively because it forms channels in powder and clumps flowers. For all other botanical materials, steam distillation, which offers the choice of high or low pressure steam, is the method of choice. Water and steam distillation offers many of the advantages of steam distillation but restricts the choice to low pressure steam. Steam distillation causes less hydrolysis of oil components, is much quicker and gives a better yield because less high boiling and water soluble components remain in the still. Steam distillation also removes the need to return the distillation waters to the still (cohobation).

In all these types of distillation the essential oil and water vapours are condensed in a condenser and collected in an oil separator. Most essential oils are lighter than water and will form a layer at the top of the separator. Much more water is distilled over than essential oil, so it is vital to remove the excess water continuously. Some oils are heavier than water and the function of the separator must be reversed.

2.2.3   Other Distillation Methods

Dry distillation is only suitable for a small range of essential oils and is often used to distil the oil from exudates such as balsams. Destructive distillation involves the formation of chemicals not present in the original botanical. A typical example is cade oil, which is produced by the destructive distillation of juniper wood. The natural status and safety in use of such oils are open to considerable question.

2.2.4   Expression of Oils

Citrus oils are often isolated from the peel by expression or 'cold pressing'. This process involves the abrasion of peel and the removal of the oil in an aqueous emulsion which is subsequently separated in a centrifuge. Expressed citrus oils have superior odour characteristics compared with distilled oils, because of the absence of heat during processing and the presence of components that would not be volatile in steam. They are also more stable to oxidation because of the presence of natural anti-oxidants, such as tocopherol, which are not volatile in steam. The lack of heat damage to the oil is also significant.

2.2.5   Extraction (see Chapter 3)

Extractions of the oils with supercritical or liquid carbon dioxide provides the advantages of a cold process and the incorporation of some of the non-volatile components. It is expensive in terms of plant and, in some cases, results in an unusual balance of extracted oil components. Carbon dioxide extracts resemble essential oils more closely than oleoresins and, when well balanced, can offer unique raw materials for flavour creation.

2.3   Further Processing of Essential Oils

Raw essential oils produced by any of these methods may be processed further in a number of ways. Simple redistillation is often carried out to clean the oil. Cassia oil, for example, is often contaminated with impurities such as iron in its raw state. Redistillation, without any separation of fractions, solves the problem.

Rectification takes this process one step further and involves the selection of desirable fractions. Some of the first fractions of peppermint oil have undesirable harsh and vegetal odours. The later fractions have heavy cloying odours. Selection of the 80–95% of oil between these extremes results in a cleaner, sweeter smelling peppermint oil.

2.3.1 Rectification

The process of rectification can be extended to cover the removal of a substantial part of the terpene hydrocarbons in an oil. These chemicals are generally more volatile than the rest of the oil and can easily be separated by distillation under vacuum. Depending on the quantity of terpene hydrocarbons originally present, the residues in the still may be as little as a thirtieth of the original volume of raw oil. The disadvantages of this process are the effects of heat and the loss of desirable components with the hydrocarbons. The profile of the oil will be changed and the degree of flavour concentration will not match the physical concentration.

Terpeneless oils are produced by distilling, under vacuum, the volatile fractions of a concentrated oil from the still residues. The main added disadvantage is the loss of desirable components in the residues. Concentrated and terpeneless oils reduce the problems of oxidation of terpene hydrocarbons, particularly in citrus oils. Terpeneless oils provide clear aqueous solutions and are especially useful for flavouring soft drinks. Sesquiterpeneless oils carry the fractionation one step further and remove the sesquiterpene hydrocarbons. They offer increased strength, stability and solubility but decreased odour fidelity because of missing components.

Fractionation can be carried out for other reasons than concentration. In some instances the objective is to isolate a desirable individual component, for example, linalol from rosewood oil. Alternatively the intention may be to exclude an undesirable component from the oil, for example, pulegone from peppermint oil.

The objective of obtaining selected parts of essential oils can be achieved in other ways. Chromatography offers a relatively cold process and great selectivity, often at a high price. Carbon dioxide extraction of oils also offers a cold process but less selectivity. Both these processes can result in concentrated oils with an unusual balance of components. Extraction of the oil with a solvent, typically a mixture of an alcohol and water, results in a well-balanced flavour. This process is very widely used in the manufacture of soft drink flavours.

2.3.2 Washed Oils

'Washed oils' produced in this way retain most of the character of the original oils. The solvent mixture may be varied to remove most or all of the hydrocarbons, depending on the degree of water solubility required. Some desirable oxygenated components are lost in the oil fraction, the 'terpenes', particularly when a low level of hydrocarbons is required in the 'washed oil'. The inefficiency of this extraction process may mean that as little as 60% of the desirable flavour in the oil is finally extracted 'Washed oils' are not often used outside soft drinks and limited dairy applications because of the high dose rates which would be required. Countercurrent liquid/liquid extraction, followed by removal of solvent, produces a high quality terpeneless oil with many of the flavour characteristics of a 'washed oil'. It provides the added bonuses of more efficient extraction and much greater flavour concentration.

The terpene hydrocarbon fractions which are the by-products of these processes are also used in the flavour industry. Quality, and usefulness, will vary considerably. Citrus terpenes from the production of 'washed oils' often retain much of the character of the original oils. They can be washed to remove residual solvent, dried, and used in the production of inexpensive nature identical flavour oils for confectionery use. Most terpenes can be used to dilute genuine oils and this is one of the most difficult types of adulteration to detect. The only change in the analysis of the oil may be in the ratios of some of the components. Distilled terpenes, particularly citrus terpenes, oxidise more readily than the original oil. Adulteration of oils with terpenes results in reduced stability as well as an inferior profile.

Some delicate raw materials are processed in a different way for use in flavours and fragrances. Flowers, such as jasmin, are first extracted with hexane. The solvent is then removed under vacuum resulting in a semi-solid mass known as a 'concrete'. This raw material is used in fragrances but is not normally soluble enough for flavour use. Extraction of the concrete with ethanol, followed by solvent removal produces an 'absolute' which mainly consists of the volatile components of the original flowers. Absolutes are obviously not strictly essential oils but they perform a similar function and are powerful raw materials in the flavourist's arsenal.

2.3.3   Oil Quality

Much of the quality control effort of flavour companies is dedicated to the detection of adulteration in essential oils and other natural raw materials. Traditional standards, often specified in official pharmacopeae, are sadly out of date. Most physical and simple chemical characteristics are easily altered to meet a specification. Gas chromatography presents a much more formidable challenge. It is very difficult to reformulate an oil in the minute detail that is revealed by this technique. One traditional method of quality control also works well. A trained nose can recognise many faults and foreign components in essential oils.

All essential oils are subject to deterioration, mainly caused by oxidation, polymerisation and hydrolysis, on storage. They should be stored dry, in full airtight containers, under nitrogen if possible, cold and away from light.

2.4   The Uses of Essential Oils

The uses of essential oils in flavours can be grouped into two broad categories. The main use for most oils is to give their own characteristic flavour to an end product. The flavour may be simple or part of a blend with other essential oils. In some cases the character may be enhanced by the addition of nature identical raw materials. The addition of ethyl butyrate (fruity, juicy) and cis-3-hexenol (green, leaf) give a typically orange juice character to orange peel oil.

The most challenging use of essential oils is in the creation of natural flavourings. Many natural flavourings cannot be produced effectively from the named product because of problems of availability, price, concentration, heat stability and variability. Essential oils sometimes contain key characteristics of other flavours and can be used, often in trace quantities, in natural flavours to enhance those characteristics. Coriander oil contains linalol, which is an important component of the natural aroma of apricots, and coriander oil is frequently used to good effect in natural apricot flavours.

The main disadvantage of this use of essential oils is that other, less desirable, components of the oil can detract from the overall character of the flavour. This problem can be reduced by the use of isolated fractions from the oil. Eugenol, derived from clove leaf oil, is almost universally used in natural banana flavours.

A small, but growing, category of essential oils find their main use in natural flavourings. The most important example is davana oil (*Artemesia pallens*, family Compositae). Only about two tons of this oil are produced annually in India but it plays a major part in the juicy character of many natural fruit flavours, especially raspberry.

2.5   The Composition of Essential Oils

The following essential oils have been selected from many hundreds. They represent the major commercial oils that are used in flavourings. Only the most important information on composition is given. Chemical formulae of the key components are shown in figure 2.2. The quantities are typical but individual oils will often vary widely in practice.

The oils belong to the following plant families:

Gramineae	lemongrass
Labiatae	cornmint, peppermint, rosemary, spearmint
Lauraceae	cassia, cinnamon, litsea cubeba
Liliaceae	garlic
Magnoliaceae	star anise
Myristicaceae	nutmeg
Myrtaceae	clove, eucalyptus

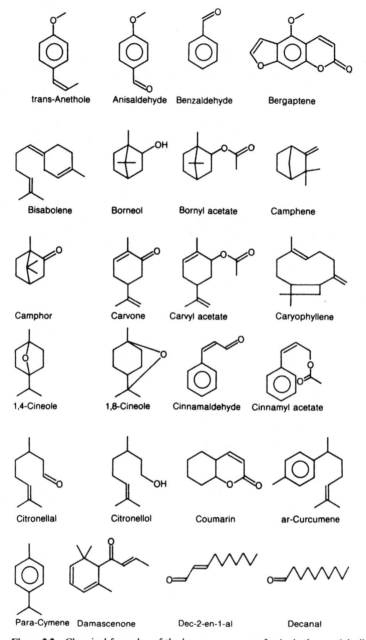

Figure 2.2 Chemical formulae of the key components of principal essential oils.

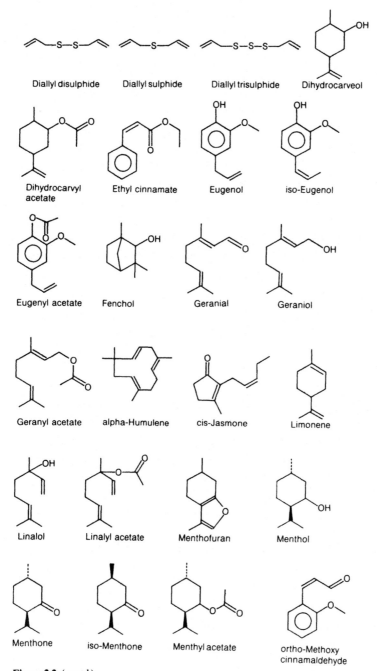

Diallyl disulphide   Diallyl sulphide   Diallyl trisulphide   Dihydrocarveol

Dihydrocarvyl acetate   Ethyl cinnamate   Eugenol   iso-Eugenol

Eugenyl acetate   Fenchol   Geranial   Geraniol

Geranyl acetate   alpha-Humulene   cis-Jasmone   Limonene

Linalol   Linalyl acetate   Menthofuran   Menthol

Menthone   iso-Menthone   Menthyl acetate   ortho-Methoxy cinnamaldehyde

Figure 2.2 (*contd.*)

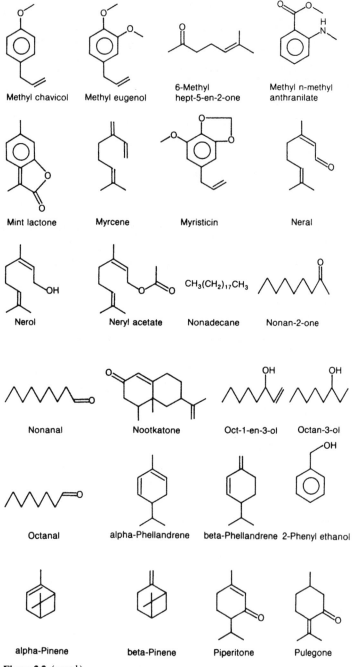

Methyl chavicol   Methyl eugenol   6-Methyl
                                     hept-5-en-2-one   Methyl n-methyl
                                                        anthranilate

Mint lactone   Myrcene   Myristicin   Neral

Nerol   Neryl acetate   CH₃(CH₂)₁₇CH₃   Nonan-2-one
                                     Nonadecane

Nonanal   Nootkatone   Oct-1-en-3-ol   Octan-3-ol

Octanal   alpha-Phellandrene   beta-Phellandrene   2-Phenyl ethanol

alpha-Pinene   beta-Pinene   Piperitone   Pulegone

Figure 2.2 (*contd.*)

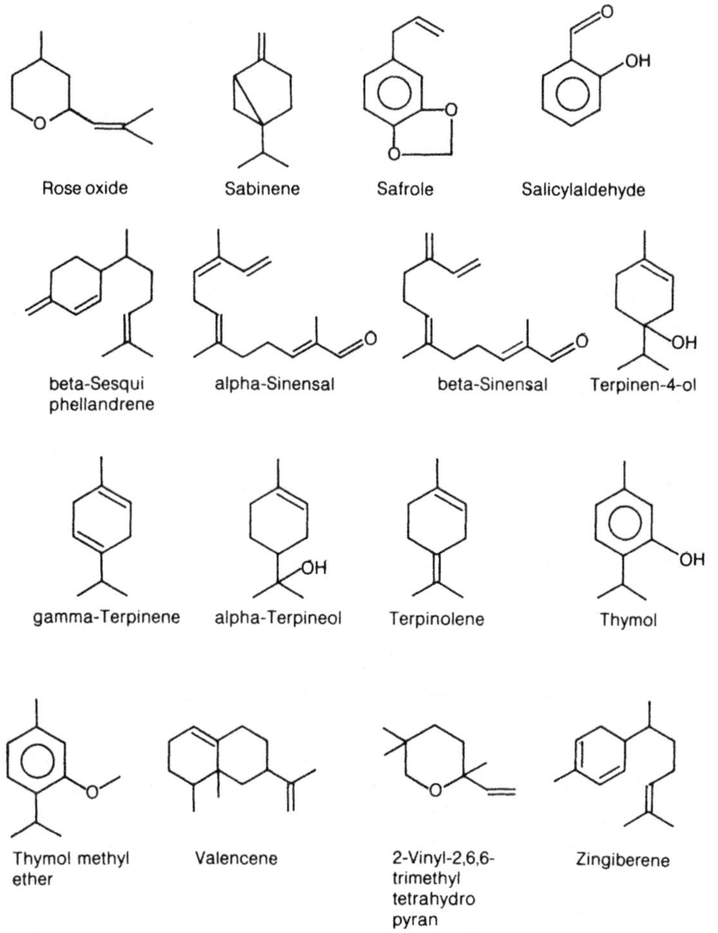

Figure 2.2 (*contd.*)

Rosaceae	rose
Rutaceae	bergamot, grapefruit, lemon, line, sweet orange, tangerine
Umbelliferae	coriander, dill
Zingiberaceae	ginger

2.5.1 Bergamot Oil

Citrus aurantium. Most of the annual total of 115 tons of bergamot oil is produced in Italy (90 tons). Other producers are the Ivory Coast (15 tons), Guinea (5 tons), Brazil (2 tons), Argentina, Spain and the U.S.S.R. [1]. Peel yields 0.5% cold pressed oil. Only small quantities of distilled oil are produced. Demand is stable. The major market is France (50%) followed by the Netherlands (17%) and the United States (15%) [2].

The major quantitative components of bergamot oil are typically:

  35%   linalyl acetate (fruity, floral, lavender)
  30%   limonene (weak, light, citrus)
  15%   linalol (light, lavender)
   7%   beta pinene (light, pine)
   6%   gamma terpinene (light, citrus, herbaceous)
   2%   geranial (lemon)
   2%   neral (lemon)

Other qualitatively important components are:

   1%   geraniol (sweet, floral, rose)
  0.4% neryl acetate (fruity, floral, rose)
  0.2% geranyl acetate (fruity, floral, rose)
  0.2% bergaptene (very weak odour)

Bergaptene is a skin sensitizer and some bergamot oil is distilled to produce bergaptene free and terpeneless oils. The level of linalol is generally higher in African than Italian oils. Adulteration of bergamot oil is sometimes carried out with synthetic linalol and linalyl acetate, together with orange and lime terpenes. It is easily detected on odour and by gas chromatography.

The major flavour use of bergamot oil is in Earl Grey tea flavours, where it is normally the major component. It is also used as a minor component in citrus soft drink flavours and some natural fruit flavours, especially apricot. There are no legal restrictions on the use of bergamot oil in flavourings. It is FEMA GRAS (2153) and Council of Europe listed.

2.5.2   Cassia Oil

Cinnamonum cassia.   Virtually all of the 160 tons of cassia oil produced annually originate in China [1]. Very small quantities are produced in Taiwan, Indonesia and Vietnam [2]. Leaves, twigs and sometimes inferior bark, yield 0.3% oil by water distillation. Demand is increasing steadily despite unpredictable production levels in China. The main markets are the United States (56%), Japan (25%) and Western Europe (11%) [2].

The major quantitative components of the oil are typically:

  85%   cinnamaldehyde (spicy, warm, cinnamon)
  11%   o-methoxy cinnamaldehyde (musty, spicy)
   6%   cinnamyl acetate (sweet, balsamic)

Other qualitatively important components are:

1%   benzaldehyde (bitter almond)
0.4% ethyl cinnamate (balsamic, fruity)
0.2% salicylaldehyde (pungent, phenolic)
0.2% coumarin (sweet, hay)

Coumarin is suspected of being toxic. Cinnamaldehyde is the most important contributor to the characteristic odour of cassia but *o*-methoxy cinnamaldehyde is mainly responsible for the unique note which distinguishes cassia from cinnamon oil. Cassia oil is often imported in a crude state and requires redistillation to improve the odour and remove metallic impurities. Adulteration of the oil with cinnamaldehyde is practised but can be easily detected by gas chromatography.

In many respects cassia was originally a 'poor man's cinnamon'. In the flavour industry, the oil makes a unique contribution in its own right. It is a major part of the traditional flavour of cola drinks. It is also used in confectionery (often called 'cinnamon'), sometimes in conjunction with capsicum oleoresin. Use of cassia oil in other natural flavours is restricted to cherry, vanilla and some nut flavours. There are no legal constraints on the use of cassia oil in flavours. It is FEMA GRAS (2258) and Council of Europe listed.

2.5.3 Cinnamon Oil

Cinnamonum zeylanicum.   Two oils are produced from the cinnamon tree. Cinnamon leaf oil is mainly produced in Sri Lanka (90 tons) together with a further 8 tons produced in India and 1 ton in the Seychelles. Only about 5 tons of cinnamon bark oil are produced annually, mainly in Sri Lanka (2 tons) [1]. Leaves yield 1% oil and bark yields 0.5% oil. Demand for the oils is declining. The main markets are Western Europe (50%) and the United States (33%) [2].

The major quantitative components of the oils are typically:

(a) Cinnamon leaf oil
    80% eugenol (strong, warm, clove)
    6%   caryophyllene (spicy, woody)
    3%   cinnamaldehyde (spicy, warm, cinnamon)
    2%   iso eugenol (sweet, carnation)
    2%   linalol (light, lavender)
    2%   cinnamyl acetate (sweet, balsamic)

(b) Cinnamon bark oil
    76% cinnamaldehyde (spicy, warm, cinnamon)
    5%   cinnamyl acetate (sweet, balsamic)
    4%   eugenol (strong, warm, clove)

    3%   caryophyllene (spicy, woody)
    2%   linalol (light, lavender)

Other qualitatively important components of the bark oil are:

    0.7%   alpha-terpineol (sweet, floral, lilac)
    0.7%   coumarin (sweet, hay)
    0.6%   1,8-cineole (fresh, eucalyptus)
    0.4%   terpinen-4-ol (strong, herbaceous)

Coumarin, as previously noted, is under suspicion of being toxic. Eugenol is the main odour component in the leaf oil but the bark oil has a complex odour with significant contributions from all the components listed above.

Cinnamon leaf oil is used in seasoning blends as an alternative to clove oil, which it resembles on odour. It is also sometimes used to adulterate the bark oil, in conjunction with cinnamaldehyde. The bark oil is used in high quality seasoning blends and occasionally in natural flavours. There are no legal constraints on the use of cinnamon leaf and bark oils. They are FEMA GRAS (2292) leaf (2291) bark and Council of Europe listed.

2.5.4 *Clove Oil*

Eugenia caryophyllata. World production of clove leaf oil is around 2000 tons, with Madagascar (900 tons), Indonesia (850 tons), Tanzania (200 tons), Sri Lanka and Brazil being the main producers. Only 10 tons of clove stem oil are produced, with Tanzania (5 tons), Madagascar (5 tons) and Sri Lanka being the major producing countries. Seventy tons of clove bud oil are produced annually. The main sources are Indonesia (40 tons), Madagascar (20 tons), with small quantities from Brazil, Sri Lanka, Tanzania, the United Kingdom and the United States [1]. Leaves and twigs yield 2% leaf oil, the stems attached to buds and flowers yield 5% stem oil and the buds yield 15% bud oil. Demand for clove oils is static with the main markets currently North America and Western Europe [2].

The major components of clove oil are typically:

    81%   eugenol (strong, warm, clove)
    15%   caryophyllene (spicy, woody)
    2%   alpha humulene (woody)
    0.5%   eugenyl acetate (warm, spicy)

Quantities of the same components in stem and bud oils are typically:

Stem	Bud	
93%	82%	eugenol
3%	7%	caryophyllene

0.3%	1%	alpha-humulene
2%	7%	eugenyl acetate

The bud oil has by far the finest odour character of the three oils but is also the most expensive. Stem oil is used as a substitute for bud oil. Leaf oil is little more than a source of natural eugenol. Adulteration of bud oil by stem, leaf oils, eugenol and stem oil terpenes is carried out but can be detected by gas chromatography.

Clove oils are used in seasoning blends but also have an interesting part to play in other natural flavours. They form an essential part of the character of banana and a useful background note in blackberry, cherry and smoke flavours. There are no legal constraints on the use of clove oils. They are FEMA GRAS (2323) bud, (2325) leaf and (2328) stem and Council of Europe listed.

2.5.5 Coriander Oil

Coriandrum sativum.   The annual production of coriander seed oil is about 100 tons, virtually all from the U.S.S.R. Other minor producers are Yugoslavia, India, Egypt, Romania, South Africa and Poland [1]. Production of coriander herb oil in France, the U.S.S.R. and Egypt is very limited. Coriander seeds yield 0.9% oil on steam distillation. The fresh herb yields only 0.02% oil. Demand is increasing slightly. The major market is the United States [2].

The major components of coriander seed oil are typically:

74%	linalol (light, lavender)
6%	gamma terpinene (light, citrus, herbaceous)
5%	camphor (fresh, camphoraceous)
3%	alpha-pinene (light, pine)
2%	para-cymene (light, citrus)
2%	limonene (weak, light, citrus)
2%	geranyl acetate (fruity, floral, rose)

The major component of coriander leaf oil is

10%	dec-2-enal (strong, orange marmalade)

Coriander seed oil can be adulterated with synthetic linalol but this is readily detectable by gas chromatography.

The seed oil is used in a very wide variety of flavour applications. It is part of the traditional flavouring of a number of alcoholic drinks, especially gin. It is widely used in meat seasonings and curry blends. It provides a very attractive natural source of linalol in natural fruit flavours, particularly apricot. The herb oil is very attractive in South Asian seasoning blends, but also provides a unique citrus character in natural flavours. There are no legal restrictions on the use of coriander oil in flavourings. It is FEMA GRAS (2334) and Council of Europe listed.

2.5.6   Cornmint Oil

Mentha arvensis.   About 3500 tons of cornmint oil (sometimes incorrectly called Chinese peppermint oil) are produced annually. It is almost all converted into menthol (1400 tons) and dementholised oil (2100 tons). China is the main producer (1400 tons of dementholised oil) followed by Brazil (400 tons), India (100 tons), Paraguay (50 tons), Taiwan (20 tons), North Korea, Thailand and Japan [1]. Dried plants yield 2.5% oil by steam distillation. Cheap synthetic menthol has reduced the demand for cornmint oil into the main markets in the United States, Western Europe and Japan [2].

The major quantitative components of the dementholised oil are typically:

    35%   laevo-menthol (cooling, light, mint)
    30%   laevo-menthone (harsh, herbal, mint)
     8%   iso-menthone (harsh, herbal, mint)
     5%   limonene (weak, light, citrus)
     3%   laevo-menthyl acetate (light, cedar, mint)
     3%   piperitone (herbal, mint)
     1%   octan-3-ol (herbal, oily)

Cornmint oil contains about 1% of pulegone (pennyroyal, mint odour) which is suspected of being toxic. The raw oils are rectified to remove some of the front and back fractions. Careful blending of fractions can reduce the characteristically harsh odour of cornmint oil but it still remains much less attractive than peppermint oil. Adulteration of cornmint oil is not a commercially attractive proposition.

Most cornmint oils are used to give a cheap peppermint flavour to a wide range of applications, often blended with true peppermint oil. It is more frequently used in blended flavours than peppermint oil because of its price advantage. Cornmint oil is Council of Europe listed with provisional limits for pulegone levels ranging from 25 ppm in food to 350 ppm in mint confectionery.

2.5.7   Dill Oil

Anethum graveolens.   The annual world production of dill weed oil is around 140 tons. The major producers are the United States (70 tons), Hungary (35 tons) and Bulgaria (20 tons), followed by the U.S.S.R. and Egypt. Annual production of dill seed oil is only 2.5 tons from the U.S.S.R, Hungary, Poland, Bulgaria and Egypt [1]. The weed oil is steam distilled from the whole plant at a yield of 0.7%. The seeds yield 3.5% oil. The major market for dill oil is the United States.

The major components of dill weed oil are typically:

    35%   dextro-carvone (warm, spearmint, caraway)
    25%   alpha-phellandrene (light, fresh, peppery)
    25%   limonene (weak, light, citrus)

Dill weed oil can be adulterated with distilled orange terpenes, but this can be readily detected by gas chromatography.

The main use for dill oil is in seasoning blends, particularly for use in pickles. There are no legal restrictions on the use of dill oils in flavourings. They are FEMA GRAS (2383) and Council of Europe listed.

2.5.8 Eucalyptus Oil

Eucalyptus globus. Cineole type eucalyptus oil production totals about 1400 tons per annum. Portugal (400 tons), South Africa (400 tons) and Spain (250 tons) are the major producers. Other producers are China (100 tons), Brazil (70 tons), Australia (60 tons), India (60 tons) and Paraguay (30 tons). An additional 400 tons of a 'eucalyptus' oil from camphor oil fractions are produced annually in China [1]. Eucalyptus leaves yield 1.5% on steam distillation. The demand for cineole type eucalyptus oils is increasing steadily. The major markets are Western Europe (60% and the United States (20%) [2].

The major components of eucalyptus oil are typically:

    75%   1,8-cineole (fresh, eucalyptus)
    10%   alpha-pinene (light, pine)
     2%   para-cymene (light, citrus)
     2%   limonene (light, weak, citrus)

Eucalyptus oils are often sold by their cineole (eucalyptol) content.

The major use for eucalyptus oil is in blends to give a fresh bright, slightly medicinal note particularly in conjunction with peppermint and aniseed oils. It may also be used in small quantities in other natural flavours such as blackcurrant. There are no legal restrictions on the use of eucalyptus oil in flavourings. It is FEMA GRAS (2466) and Council of Europe listed.

2.5.9 Garlic Oil

Allium sativum. Annual production of garlic oil is around 10 tons. The major producers are Mexico (6 tons), Italy (2 tons) and Egypt (1.5 tons), followed by

China, Bulgaria, Spain, France and the United Kingdom [1]. Garlic bulbs yield 0.1% oil on steam distillation. Demand is steady in the major markets in France and Spain [2].

The major components of garlic oil are typically:

    30%   diallyl disulphide (strong, garlic)
    30%   diallyl trisulphide (strong, heavy, garlic)
    15%   diallyl sulphide (strong, fresh, garlic)

Garlic oil can be adulterated with nature identical raw materials but this can be detected by gas chromatography.

The main use of the oil is in seasoning blends but it can also be used, in very small quantities, in other natural flavours to give a sulphur note. There are no legal restrictions on the use of garlic oil in flavourings. It is FEMA GRAS (2503) and Council of Europe listed.

2.5.10   *Ginger Oil*

Zingiber officinale.   The annual production of ginger oil is close to 55 tons. The major producers are China (20 tons) and India (20 tons) with much smaller quantities from Sri Lanka, Jamaica, Australia and South Africa. Significant quantities are produced in Europe and the United States from imported roots [1]. Dried roots yield 2% oil on steam distillation. Demand is increasing slowly. The major markets are the United States, Western Europe and Japan [2].

The major quantitative components of ginger oil are typically:

    35%   zingiberene (warm, woody)
    10%   AR-curcumene (woody)
    10%   beta-sesquiphellandrene (woody)
     8%   bisabolene (woody, balsamic)
     6%   camphene (camphoraceous, light)
     3%   beta-phellandrene (light, fresh, peppery)
     2%   1,8-cineole (fresh, eucalyptus)

Other qualitatively important components are:

    0.5%   bornyl acetate (camphoraceous, earthy, fruity)
    0.5%   linalol (light, lavender)
    0.3%   geranial (lemon)
    0.2%   neral (lemon)
    0.2%   nonan-2-one (pungent, blue cheese)
    0.1%   decanal (strong, waxy, peel)

Oils from some sources, especially Australia, contain much higher levels of geranial and neral. Terpeneless ginger oil is produced to improve solubility in soft drinks, but this product is frequently a complicated recombination of a number of different fractions from a distillation. It varies widely in strength and character. Ginger oil is not often adulterated.

Large quantities of the oil are used in soft drinks. It is also used in spice blends for bakery and confectionery. The oil lacks the hot taste character of the oleoresin and the two raw materials are often blended together. Small quantities of ginger oil may be used in natural flavours such as raspberry. There are no legal restrictions on the use of ginger oil in flavourings. It is FEMA GRAS (2522) and Council of Europe listed.

2.5.11   *Grapefruit Oil*

Citrus paradisi.   The total world production of grapefruit oil is about 180 tons. The major producer is Brazil (40 tons) followed by the United States, Israel, Argentina and New Zealand [1]. Peel yields 0.4% cold pressed oil. Demand is steady. The major market is the United States.

The major quantitative components of grapefruit oil are typically:

> 90%   limonene (weak, light, citrus)
> 2%   myrcene (light, unripe mango)

Other qualitatively important components are:

> 0.5%   octanal (strong, fresh, peel)
> 0.4%   decanal (strong, waxy, peel)
> 0.3%   linalol (light lavender)
> 0.2%   nootkatone (strong, peel, grapefruit)
> 0.1%   citronellal (strong, citrus, green)
> 0.1%   neral (lemon)
> 0.1%   geranial (lemon)
> 0.02%   beta-sinensal (strong, orange marmalade)

Grapefruit oil often smells disappointingly like orange oil, rather than grapefruit. Nootkatone is frequently taken as an indicator of quality but it only represents a part of the recognisable character. Small quantities of grapefruit essence oil are recovered during juice concentration. This oil has a more typical fresh juice character but is relatively unstable to oxidation. Concentrated and terpeneless oils, together with extracts, are produced to solve the twin problems of oxidation and solubility. They are frequently not very reminiscent of grapefruit on odour. Grapefruit oil can be adulterated with orange terpenes and, as the genuine oil has more than a passing resemblance to orange, it is easier to detect by gas chromatography than odour.

The main use of grapefruit oil in flavours is to impart a grapefruit character to a wide range of application. It is sometimes mixed with other citrus flavours but is not often used outside this field. Much more natural tasting grapefruit flavours can be achieved by adding nature identical raw materials (catty, green and juicy notes) to the oil, and most of the flavours produced are of this type. There are no legal restrictions on the use of grapefruit oil in flavourings. It is FEMA GRAS (2530) and Council of Europe listed.

2.5.12 Lemon Oil

Citrus limon.   About 2500 tons of lemon oil are produced annually. The United States (600 tons), Italy (500 tons) and Argentina (500 tons) are the major producers. Other producers are Brazil (190 tons), Ivory Coast (120 tons), Greece (100 tons), Spain (100 tons), Israel, Cyprus, Australia, Peru, Guinea, Indonesia, Venezuela and Chile [1]. Lemon peel contains 0.4% oil and most production is by cold pressing. Cheaper, but inferior, distilled oil is mainly used in the production of terpeneless oil. Small quantities of lemon juice oil are produced during juice concentration. Although different varieties of tree are planted, the main reasons for the differences in character of oils from different areas are the dissimilarities of terrain, climate and production methods. Sicilian oil has the best odour characteristics. Demand for lemon oil is only increasing slowly. The main markets are Western Europe (40%), the United States (35%) and Japan (8%) [2].

The major quantitative components of cold pressed lemon peel oil are typically (see also Appendix I):

    63%   limonene (weak, light, citrus)
    12%   beta-pinene (light, pine)
     9%   gamma-terpinene (light, citrus, herbaceous)

Other qualitatively important components are:

    1.5%   geranial (lemon)
    1.0%   neral (lemon)
    0.5%   neryl acetate (fruity, floral, rose)
    0.4%   geranyl acetate (fruity, floral, rose)
    0.2%   citronellal (strong, citrus, green)
    0.2%   linalol (light, lavender)
    0.1%   nonanal (strong, peel)

The citral (neral and geranial) level in United States oil is lower than normal and these oils have an individual, rather atypical odour. Winter oils are better than those produced in summer and generally command about 15% higher prices. Production methods also affect the quality of the oil. Sfumatrice oils have a fine fresh lemon character and change in colour from yellow/green to yellow/orange as the fruit ripens. They are on average 15% more expensive than pelatrice oils, which have a poorer, grassy odour and change in colour from dark green/yellow to yellow brown through the season. The terpene hydrocarbons which constitute the bulk of the oil are insoluble in water and susceptible to oxidation. Extraction, concentration and deterpenation are carried out to produce stable, soluble lemon flavours for soft drinks. Terpeneless lemon oils give a reasonably balanced lemon character. Adulteration of cold pressed lemon oil with distilled oil and terpenes is detectable by gas chromatography.

Lemon oils are used in a wide variety of applications. Juice oil is not widely used because it does not have the attractive fresh character of orange juice oil. Nature identical components can be added to duplicate the elusive note of fresh juice. Lemon oil is also used in other flavours, such as butterscotch, pineapple and banana. There are no legal constraints on the use of lemon oil in flavourings. It is FEMA GRAS (2625) and Council of Europe listed.

2.5.13 Lemongrass Oil

Cymbopogon flexuosus and *citratus*.   *Cymbopogon flexuosus* is known as East Indian lemongrass and is indigenous to South Africa. *Cymbopogon citratus* is mainly cultivated in Central and South America and is known as West Indian lemongrass. Only 310 tons are still produced. India (120 tons) and China (120 tons) are the major producers, followed by Guatemala (40 tons), Brazil (25 tons), Sri Lanka, Haiti and the U.S.S.R. [1]. Dried grass yields 0.5% oil on steam distillation. Demand continues to decline because of the increasing dominance of *Litsea cubeba* as a source of natural citral. The main markets are Western Europe (30%), the United States (20%) and the Soviet Union (20%) [2].

The major quantitative components of East Indian lemongrass oil are typically:

40% geranial (lemon)
30% neral (lemon)
7% geraniol (sweet, floral, rose)
4% geranyl acetate (fruity, floral, rose)
2% limonene (light, weak, citrus)

Other qualitatively important components are:

1.4% caryophyllene (spicy, woody)
1.1% 6-methyl hept-5-en-2-one (pungent, fruity, herbaceous)
0.9% linalol (light, lavender)
0.5% citronellal (strong, citrus, green)

Before the advent of *Litsea cubeba*, lemongrass oil was the main source of natural citral. The other components of the oil also have a strongly characteristic effect which is evident even in citral fractionated from lemongrass oil. Adulteration with synthetic citral can be detected by gas chromatography.

Lemongrass citral still has some use in lemon flavours. Some traditional lemonade flavours derived part of their recognisable character from lemongrass citral and this can be recaptured by using a little lemongrass oil together with synthetic citral. There are no legal constraints on the use of lemongrass oil. It is FEMA GRAS (2624) and Council of Europe listed.

2.5.14 Lime Oil

Citrus aurantifolia.   The annual production of distilled lime oil is around 450 tons. Mexico (180 tons), Peru (130 tons) and Haiti (50 tons) are the major producers, followed by Brazil (25 tons), Cuba (25 tons), Ivory Coast (12 tons), the Dominican Republic (6 tons), Guatemala (5 tons), Jamaica (5 tons), Ghana, Swaziland and China. Only 160 tons of cold pressed oil are produced each year in Brazil (90 tons), the United States (40 tons), Mexico (25 tons) and Jamaica. Ten tons of oil are produced annually from a related variety, Persian lime (*Citrus latifolia*) in Brazil [1]. Lime peel yields 0.1% of oil. Demand for lime oil continues to increase. The major market is the United States (67%) followed by the United Kingdom (10%) [2].

The major quantitative components of distilled lime oil are typically:

52%	limonene (light, weak, citrus)
8%	gamma-terpinene (light, citrus, herbaceous)
7%	alpha-terpineol (sweet, floral, lilac)
5%	terpinolene (fresh, pine)
5%	para-cymene (light, citrus)
3%	1,4-cineole (fresh, eucalyptus)
2%	1,8-cineole (fresh, eucalyptus)
2%	beta-pinene (light, pine)

Other qualitatively important components are:

1%	bisabolene (woody, balsamic)
0.7%	fenchol (camphoraceous, earthy, lime)
0.7%	terpinen-4-ol (strong, herbaceous)
0.5%	borneol (camphoraceous, earthy, pine)
0.4%	2-vinyl-2,6,6-trimethyl tetrahydropyran, (light, lime)

The major quantitative components of cold pressed lime oil are:

45%	limonene (light, weak, citrus)
14%	beta-pinene (light, pine)

8%   gamma-terpinene (light, citrus, herbaceous)
3%   geranial (lemon)
2%   neral (lemon)

The distilled oil is chemically very different from the cold pressed oil because it is distilled over acid. Much of the citral (neral and geranial) is lost in the process and with it the raw/fresh lemon character which typifies lemon peel. This character is replaced in the distilled oil by the lilac/pine/camphoraceous/earthy complex character which is normally recognised as lime flavour. Terpeneless lime oil is produced to improve solubility in soft drinks. Adulteration of lime oil is normally carried out by adding synthetic terpineol, terpinolene and other components to lime terpenes. It can often be easily identified on odour and can also be detected by gas chromatography.

Distilled lime oil is a major component of flavours for cola drinks. It is also used in traditional clear lemon lime soft drinks. Cold pressed oil is used in some more recent lemon lime drinks. Lime is little used in other natural flavours. There are no legal constraints on the use of lime oils in flavourings. Lime oil is FEMA GRAS (2631) and Council of Europe listed.

2.5.15   Litsea Cubeba Oil

Litsea cubeba.   About 900 tons of oil are produced in China [1]. The fruit yields about 3.2% oil on steam distillation. Demand for the oil is increasing rapidly. The main market is China itself, followed by the United States, Western Europe and Japan.

The major quantitative components of the oil are typically:

41%   geranial (lemon)
34%   neral (lemon)
 8%   limonene (weak, light, citrus)
 4%   6-methyl hept-5-en-2-one (pungent, fruity, herbaceous)
 3%   myrcene (light, unripe mango)
 2%   linalyl acetate (fruity, floral, lavender)
 2%   linalol (light, lavender)

Other qualitatively important components are:

 1%   geraniol (sweet, floral, rose)
 1%   nerol (sweet, floral, rose)
0.5%   caryophyllene (spicy, woody)
0.5%   citronellal (strong, citrus, green)

The dominant citral (neral and geranial) odour is considerably modified by the other components of the oil, particularly methyl heptenone and citronellal.

The oil is frequently fractionated to increase the citral content above 95% and improve the purity of the lemon odour. It is possible to treat the oil with bisulphite to form bisulphite addition compounds of the aldehydes. The original aldehydes can later be regenerated, giving a high citral content and an attractive odour. This process is chemical and the purified oil should not be described as 'natural'. Adulteration with synthetic citral can be detected by gas chromatography.

Citral redistilled from *Litsea cubeba* oil is widely used in lemon flavours. It demonstrates good stability in end-products and many of the minor impurities reflect minor components of lemon oil itself. It can also be used in some natural fruit flavours, especially banana. *Litsea cubeba* oil is Council of Europe listed.

2.5.16   *Nutmeg Oil*

Myristica fragrans.   Annual production of nutmeg oil is around 180 tons. The major producer is Indonesia (120 tons), followed by Sri Lanka (30 tons), India (5 tons). Some European oil is distilled from Grenadian nutmegs [1]. Nutmegs yield 11% oil on steam distillation. Mace, the coating of the nutmeg, yields 12% oil but very little is produced. Demand for nutmeg oil is static. By far the main market is the United States (75%), followed by Western Europe [2].

The major components of nutmeg oil are typically:

22%   sabinene (light, peppery, herbaceous)
21%   alpha-pinene (light, pine)
12%   beta-pinene (light, pine)
10%   myristicin (warm, woody, balsamic)
 8%   terpinen-4-ol (strong, herbaceous)
 4%   gamma-terpinene (light, citrus, herbaceous)
 3%   myrcene (light, unripe mango)
 3%   limonene (weak, light, citrus)
 3%   1,8-cineole (fresh, eucalyptus)
 2%   safrole (warm, sweet, sassafras)

Safrole and myristicin are suspected of being toxic. Myristicin is normally present at a much lower level in oil distilled from West Indian nutmegs, but unfortunately this is not a normal item of commerce. Terpeneless nutmeg oil is produced to improve solubility for use in soft drinks. The oil can be adulterated by the addition of terpenes and nature identical raw materials, but this is easily detected on odour and by gas chromatography.

Nutmeg oil is a major component of cola flavours, and this use accounts for most of the world-wide production. It is also used in meat seasonings and in spice mixtures for bakery products. Small quantities can be used in natural fruit flavours, where it imparts depth and richness. Nutmeg oil is FEMA GRAS (2793) listed together with mace oil FEMA GRAS (2653). Both oils are

Council of Europe listed with provisional limits for safrole levels ranging from less than 1 ppm in food to 15 ppm in products 'containing mace or nutmeg'.

2.5.17 Sweet Orange Oil

Citrus sinensis. Over 16 000 tons of orange oil are produced each year. The major producers are Brazil (10 000 tons) and the United States (5000 tons). Other producers are Israel (400 tons), Italy (300 tons), Australia (50 tons), Argentina, Morocco, Spain, Zimbabwe, Cyprus, Greece, Guinea, the U.S.S.R, South Africa, Indonesia and Belize [1]. The majority of the oils are cold pressed from the peel of the fruit. Only small quantities are still produced by steam distillation. A different type of oil is produced during the concentration of orange juice. Orange juice or essence oil is the oil phase of the volatiles recovered during the process. Typical yields from the fruit are 0.28% of peel oil and 0.008% of juice oil [3]. Different varieties of tree are planted to ripen at different times of year, hence the terms early, mid and late season. Late season oils are said to be the best and are usually produced from the variety 'Valencia'. Demand for orange oil is increasing steadily. The main markets are the United States (30%), Western Europe (30%) and Japan (20%) [2].

The major quantitative components of cold pressed sweet orange oil are typically (see also Appendix I):

94%   limonene (weak, light, citrus)
 2%   myrcene (light, unripe mango)

Other qualitatively important components are:

0.5%   linalol (light, lavender)
0.5%   octanal (strong, fresh, peel)
0.4%   decanal (strong, waxy, peel)
0.1%   citronellal (strong, citrus, green)
0.1%   neral (lemon)
0.1%   geranial (lemon)
0.05%   valencene (orange juice)
0.02%   beta-sinensal (strong, orange marmalade)
0.01%   alpha-sinensal (strong, orange marmalade)

The total aldehydes in orange oil have traditionally been taken as a quality indicator. Brazilian oils often have fairly low aldehyde contents. At the other end of the scale Zimbabwean oils may contain double the average level of aldehydes. Aldehyde content is an important factor in determining the yield of terpeneless oils but it is not a reliable guide to quality. Minor components, particularly the sinensals make a major contribution.

Some countries produce oils with different odour characteristics which command high prices. Sicilian and Spanish oils are the most important examples. Orange juice oil has a fresher, juicy odour. It contains much more valencene (up to 2%) together with traces of ethyl butyrate, hexanal and other components which are responsible for the fresh odour of orange juice. The major component, d-limonene, has little odour value and is susceptible to oxidation giving rise to unpleasant off notes. It is also insoluble in water at normal use levels. Some orange oil is processed to reduce the d-limonene content. Concentrated and terpeneless oils produced by distillation have 'flatter', less fresh odours because octanal and other volatile components are removed with the limonene. Octanal can be recovered and added back to improve odour quality. Extraction and chromatography produce better results. Orange terpenes are the by-product of all these processes and find a ready market as a solvent for flavour and fragrances.

Adulteration of orange oil is an infrequent practice because the market price of the oils from the major producers is so low. Higher priced oils are sometimes diluted with oils from cheaper sources. This practice can be very difficult to detect.

Orange oil gives an acceptable orange flavour in many applications. A proportion of juice oil may be added to give a fresh effect. In nature identical flavours orange oil will often provide a base. Components found in orange juice could be included to duplicate the fresh juice taste. Other attractive top notes which can be subtly emphasised in orange oil based flavours include violet, lemon and vanilla. Orange oil is also used in other flavours, particularly apricot, peach, mango and pineapple. There are no legal constraints on the use of orange oil in flavours. It is FEMA GRAS (2821) and Council of Europe listed.

2.5.18 Peppermint Oil

Mentha piperita. The majority of the 2200 tons of peppermint oil produced annually is from the United States (2000 tons). Other producers are the U.S.S.R., France, South Africa, Italy, Morocco, Yugoslavia, Hungary and Bulgaria [1]. The oils are produced by steam distillation at a yield of 0.4%. Demand for peppermint oil is only increasing slowly. The main market is the United States [2].

The major quantitative components are typically:

50%	laevo-menthol (cooling, light, mint)
20%	laevo-menthone (harsh, herbal, mint)
7%	laevo-menthyl acetate (light, cedar, mint)
5%	1,8-cineole (fresh, eucalyptus)
4%	menthofuran (sweet, hay, mint)
3%	limonene (weak, light, citrus)
3%	iso-menthone (harsh, herbal, mint)

Other qualitatively important components are:

0.4%   octan-3-ol (herbal, oily)
0.1%   oct-1-en-3-ol (raw, mushroom)
0.03%   mint lactone (creamy, sweet, mint)

Peppermint oil contains about 1% of pulegone (pennyroyal, mint odour) which is suspected of being toxic. Menthofuran is the main analytical difference between peppermint oil and the cheaper, but similar, cornmint oil, but it is not a reliable indicator of quality. Raw oils are usually rectified to remove some of the front and back fractions. The nature of this processing varies considerably and can contribute to a recognisable house style. Oils with heavier cuts are sometimes described as 'terpeneless' and can be used for cordials or liqueurs where a clear end-product is required. Complicated re-blending of individual fractions from peppermint oil can achieve selective results, reduction in the level of pulegone or softening of the odour profile.

Adulteration of peppermint oil with cornmint oil and terpenes occurs and can be easier to detect on odour than by analysis. Stretching with nature identical components is much more unusual and is rarely effective.

The main use of peppermint oil is to give a peppermint flavour to a wide range of applications. It is frequently blended with spearmint oils and also with many other common flavourings, including eucalyptus oil, methyl salicylate and anethole. Small quantities can be used to give subtle effects in natural fruit flavours. Peppermint oil is FEMA GRAS (2848) and Council of Europe listed with provisional limits for pulegone levels ranging from 25 ppm in food to 350 ppm in mint confectionery.

2.5.19   *Rose Oil*

Rosa damescena.   Total production of rose oil and concrete from *Rosa damascena* and *Rosa centifolia* (Rose de Mai) is around 15 tons. The major producers are Turkey (4 tons), Bulgaria (3 tons) and the U.S.S.R. (2 tons) (all *damascena*) and Morocco (3 tons) (*centifolia*). Other producers are India, Saudia Arabia, France, South Africa and Egypt [1]. Flowers yield 0.02% oil (sometimes called otto) on steam distillation or 0.2% concrete on extraction with hexane. The concrete yields about 50% absolute. Demand is increasing. The major markets are the United States and Western Europe [2].

The major quantitative components of rose oil are typically:

50%   citronellol (fresh, floral, rose)
18%   geraniol (sweet, floral, rose)
10%   nonadecane (odourless)
 9%   nerol (sweet, floral, rose)

2%   methyl eugenol (warm, musty)
2%   geranyl acetate (fruity, floral, rose)

Other qualitatively important components are:

1%   eugenol (strong, warm, clove)
1%   2-phenyl ethanol (sweet, honey, rose)
0.6%   rose oxide (warm, herbaceous)
0.5%   linalol (light, lavender)
0.04% damascenone (strong, berry, damson)

The volatile fractions of concretes and absolutes contain a much higher level (up to 60%) of 2-phenyl ethanol. The relatively low level in distilled oil is caused by the fact that 2-phenyl ethanol is not steam volatile. Nonadecane and other stearoptenes in the distilled oil contribute very little to the odour of the oil but they solidify in the cold and cause solubility problems. Oils with these components removed are called stearopteneless. Rose oil is frequently adulterated with nature identical components such as citronellol and geraniol. This is usually obvious on odour and can be detected by gas chromatography.

The oil is used alone to flavour Turkish Delight and many other products in Asia. It is a very useful raw material in other natural flavours. Small quantities can be used to very good effect in many fruit flavours, especially raspberry. There are no legal restrictions on the use of rose oil in flavourings. It is FEMA GRAS (2989) and Council of Europe listed.

2.5.20   Rosemary Oil

Rosmarinus officinalis.   About 250 tons of rosemary oil are produced each year. The major producer is Spain (130 tons), followed by Morocco (60 tons), Tunisia (50 tons), India, the U.S.S.R., Yugoslavia, Portugal and Turkey [1]. Fresh plants yield 0.5% steam distilled oil. Demand for rosemary oil continues to decline steadily. The major market is Western Europe.

The major components of rosemary oil are typically:

20%   alpha-pinene (light, pine)
20%   1,8-cineole (fresh, eucalyptus)
18%   camphor (fresh, camphoraceous)
7%   camphene (camphoraceous, light)
6%   beta-pinene (light, pine)
5%   borneol (camphoraceous, earthy, pine)
5%   myrcene (light, unripe mango)
3%   bornyl acetate (camphoraceous, earthy, fruity)
2%   alpha-terpineol (sweet, floral, lilac)

Moroccan and Tunisian oils normally contain about 40% of 1,8-cineole, with correspondingly lower levels of alpha-pinene, camphor and camphene. Camphor and eucalyptus fractions are used to adulterate rosemary oil and can easily be detected by gas chromatography.

Rosemary oil is used in seasoning blends and in flavour oil blends with a medicinal connotation. It is not often used in other natural flavours. There are no restriction on the use of rosemary oil in flavourings. It is FEMA GRAS (2992) and Council of Europe listed.

2.5.21 Spearmint Oil

Mentha spicata.   One thousand tons out of the total world production of 1400 tons are produced in the United States. Other producers are China (300 tons), Italy, Brazil, Japan, France and South Africa [1]. About 80% of the oil produced in the United States is from *Mentha spicata* and is called 'native' oil. The less hardy, but superior in terms of odour quality, *Mentha cardiaca* cross produces an oil called 'Scotch'. Western Europe is the largest market with demand increasing very slowly [2].

The major quantitative components are typically:

    70%   laevo-carvone (warm, spearmint)
    13%   limonene (weak, light, citrus)
     2%   myrcene (light, unripe mango)
     2%   1,8-cineole (fresh, encalyptus)
     2%   carvyl acetate (sweet, fresh, spearmint)

Other qualitatively important components are:

     1%   dihydrocarvyl acetate (sweet, fresh, spearmint)
     1%   dihydrocarveol (woody, mint)
     1%   octan-3-ol (herbal, oily)
    0.4% *cis*-jasmone (warm, herbal)

The raw oil is rectified to reduce the harsh sulphurous front fractions and sometimes to reduce the limonene producing a 'terpeneless' oil. The odour character of spearmint oil frequently improves considerably with age. Adulteration with laevo-carvone and blending from different sources occurs but can be detected on odour and by analysis.

The main use of spearmint oil is to give a spearmint flavour to chewing gum and oral hygiene flavours. It is often blended with peppermint and other flavours. There are no restrictions on the use of spearmint oil in flavours. It is FEMA GRAS (3032) and Council of Europe listed.

2.5.22   *Star Anise Oil*

Illicium verum.   About 90 tons of star anise oil are produced each year. China is the major producer (70 tons) and the oil is frequently called 'China star anise oil'. Other producers are Vietnam (15 tons), North Korea and the U.S.S.R. [1]. Dried fruit yields 8.5% oil on steam distillation. This oil should not be confused with the chemically similar, but botanically unrelated, anise seed oil, which is produced from *Pimpinella anisum*, an umbelliferous plant. Only 8 tons of anise seed oil are still produced each year. Demand for star anise oil continues to decline. France is the main market [2].

The major components of star anise oil are typically:

    87%   *trans*-anethole (strong, sweet, anise)
     8%   limonene (weak, light, citrus)

Other qualitatively important components are:

     1%   anisaldehyde (sweet, floral, hawthorn)
   0.8%   linalol (light, lavender)
   0.5%   methyl chavicol (strong, sweet, tarragon)

Redistilled oils, containing less limonene, are prepared for use in clear drinks. The oil can be adulterated with anethole but this can be detected by gas chromatography.

Star anise oil is used in large quantities in alcoholic drinks. It is also a popular flavouring in confectionery, particularly with a medicinal connotation, and in oral hygiene applications. Use in other natural flavours is restricted to small quantities in flavours such as cherry. There are no legal restrictions on the use of star anise oil in flavourings. It is FEMA GRAS (2096) and Council of Europe listed.

2.5.23   *Tangerine Oil*

Citrus reticulata.   World production of tangerine oil is about 300 tons. The main producers are Brazil (250 tons) and the United States (45 tons), followed by the U.S.S.R., South Africa and Spain. Production of the closely related mandarin oil is 120 tons, from Italy (50 tons), China (40 tons), Argentina (10 tons), Brazil (10 tons), the Ivory Coast, the United States and Spain [1]. Cold pressed peel yields 0.5% oil. Demand is increasing steadily in the main markets of the United States, Western Europe and Japan.

Mandarin oil has a strong aromatic/musty odour and justifiably commands a much higher price than tangerine oil, which is more reminiscent of orange.

The major quantitative components of tangerine oil are typically:

93% limonene (weak, light, citrus)
 2% gamma-terpinene (light, citrus, herbaceous)
 2% myrcene (light, unripe mango)

Other qualitatively important components are:

0.7%   linalol (light, lavender)
0.4%   decanal (strong, waxy, peel)
0.3%   octanal (strong, fresh, peel)
0.1%   thymol methyl ether (sweet, thyme, herbaceous)
0.05% alpha-sinensal (strong, orange marmalade)
0.03% thymol (sweet, phenolic, thyme)

The major quantitative components of mandarin oil are typically:

72% limonene (light, weak, citrus)
18% gamma-terpinene (light, citrus, herbaceous)

Other qualitatively important components are:

0.8%   methyl *n*-methyl anthranilate (heavy, musty, orange blossom)
0.5%   linalol (light, lavender)
0.2%   decanal (strong, waxy, peel)
0.1%   thymol (sweet, phenolic, thyme)
0.05% alpha-sinensal (strong, orange marmalade)

Concentrated and terpeneless oils are produced, together with extracts, to give flavours used in clear soft drinks. Much of the character of the original oil is retained in terpeneless oils. Adulteration is normally carried out by adding synthetics, such as methyl *n*-methyl anthranilate, to sweet orange oil. It is usually easy to detect by gas chromatography.

Mandarin and tangerine oils are widely used in soft drink flavours and confectionery both alone and in conjunction with orange oils. They also find good use in other natural fruit flavours, especially mango and apricot. There are no legal restrictions on the use of either oil in flavourings. They are FEMA GRAS (3041) tangerine, (2657) mandarin and Council of Europe listed.

3 Oleoresins, tinctures and extracts

D.A. MOYLER

3.1 Introduction

3.1.1 General Comments

When preparing or choosing a natural ingredient for incorporation into a food or beverage product, some initial thought about its form or the isolation method used to obtain the flavour principles can save considerable time at later stages of the development of consumer products. This simple principle applies whether preparing vegetables in the kitchen before making a sauce or choosing one of a spectrum of ginger extracts for making a flavouring suitable for ginger ale. This chapter illustrates the principles of extraction in the food industry from a practical not a theoretical view point. Illustrative examples of actual problems are explained.

The form of the finished product will influence the selection process. For instance, a clear carbonated ginger ale requires a water soluble ginger extract whereas a cloudy carbonated 'brewed style' ginger beer does not. For the clear beverage, a soluble ginger CO_2 extract would be recommended (not the outdated steam distilled and oleoresin ginger mixture which precipitates a sticky resinous deposit that has to be filtered out). A cloudy ginger beverage can be made by boiling ginger root in water to give flavour and cloud together.

3.1.2 Costs

Financial constraints of formulating products are not just those of raw materials. It is pointless using cheaper raw materials if the processing or labour cost in rendering them to a stable soluble form in the product, escalate the total cost to one higher than that of a selective soluble extract that could be incorporated by simple mixing. Cost efficiency and the accurate material and process costings which make up that total are the key to successful product formulating in today's competitive world.

3.1.3 Raw Materials and Processes

If simple chopped or powdered dry plant materials can be used to give the required flavour to a slowly cooked traditional curry dish, then use them.

However, if an instant convenience curry sauce is required for some customers, then extracts of plant materials are preferred, as they quickly and consistently disperse into the sauce.

When the instant convenience of a carbonated cola beverage is required, many processing and extraction technologies are needed to prepare the vegetable and spice ingredients into a drinkable form, e.g. roasting, grinding and extraction of kola nuts; cold expression and subsequent solubilisation of citrus fruit oils; grinding and distillation of spices such as cassia bark, nutmeg kernels, ginger root, peppercorns, etc. All this activity is before even thinking about the caramelisation of sugar, addition of acid, water quality and that essential sparkle provided by carbon dioxide.

These technologies are so much a part of our lives that their costs of manufacture are accepted. To clarify the differences between forms of natural

Table 3.1 Some terminology used in the preparation of oleoresins, tinctures and extracts [1]

Type	Definition	Example
Concrete	Extract of previously live plant tissue; contains all hydrocarbons, soluble matter and are usually solid, waxy substances	All flowers (jasmin, orange flower, rose, carnation etc.)
Absolute	An alcohol extract of a concrete which eliminates waxes, terpene and sesquiterpene hydrocarbons	All flowers (jasmin, clove, cubeb, cumin, lavender, chamomile, etc.)
Balsam	A natural raw material which exudes from a tree or plant; they have a high content of benzoic acid, benzoates and cinnamates	Peru, tolu, canada, fir needle, etc.
Resin	They are either natural or prepared; natural resins are exudations from trees or plants, and are formed in nature by oxidation of terpenes; prepared resins are oleoresins from which the essential oils have been removed	Orris, olibanum, mastic, cypress and flowers
Oleoresin	Natural oleoresins are exudations from plants; prepared oleoresins are liquid extracts of plants yielding the oleoresin upon evaporation	Gurjun, lovage, onion, pepper, melilotus, etc.
Extract	Concentrated products obtained from solvent treatment of a natural product	St. John's bread, rhatany root, mate, fenugreek, etc.

botanical extracts, Table 3.1 lists some of the terminology used within the industry.

3.2   Plant Materials

3.2.1   *Origin*

To ensure consistency in a formulated product containing natural extracts, it is important to specify the country of origin of a plant material ingredient as well as its botanical name. For instance, nutmeg kernels (*Myristica fragrans*) from the East Indies (E.I.) have a different organoleptic profile to those from the West Indies (W.I.). The E.I. nutmeg has a higher myristicin level (5% compared to 4%) which is stronger for flavour use [2] where nutmeg from the W.I. is low in safrole (0.2% compared to 3%), which is considered desirable for safety reasons [3]. Sri Lankan nutmegs are different again [4].

Such examples of variations in plants, soil, climate and growing conditions are commonplace in the essential oils and extract industry. The fact that a supplier can give details of botany, origin and crop time is usually a sign that they are thoroughly conversant with the material and it is less likely to be blended (with other origin naturals or different plant materials or even nature identical chemicals).

Plant materials from different origins are also valued on quality and some, which are considered from a superior origin, are at a premium price in the market place. Purchasers should always be on their guard against inferior qualities being passed off as something they are not. Periods of market shortages are the most usual time for the 'relaxation of standards'. However, most suppliers with a high reputation will supply material from a specific origin on request, or offer a 'blend' to meet a specification within the price constraints applicable to today's competitive products.

3.2.2   *Crop to Crop Variations*

Sometimes, despite all the skill that a grower and processer can apply, mother nature provides plants which, when extracted, are outside the previously encountered range of characteristics. This does not usually present a serious problem as some blending with previous carry over crops or oils from nearby growing areas can take place. However, if oils are being marketed as being 100% natural and from a named specific origin, due allowance should be made for a possible wider variation than would otherwise be encountered as due to 'seasonal variations'.

One example is cinnamon bark oil which contains a large proportion of cinnamaldehyde. The leaf oil, obtained from the same tree, is principally eugenol. The U.S. Essential Oil Association has published a specification for the bark oil as 55–75% cinnamaldehyde.

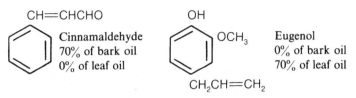

Cinnamaldehyde
70% of bark oil
0% of leaf oil

Eugenol
0% of bark oil
70% of leaf oil

By careful raw material selection and careful low temperature processing, it is possible routinely to obtain bark oils with a cinnamaldehyde content in excess of 80%.

The bark oil from Sri Lanka (a major source) is frequently 'standardised' to a minimum content of cinnamaldehyde, namely two grades at 60% and 30%. Cinnamon leaf oil is used for this purpose. This dilution serves to even out crop to crop and botanical variations. In any product application where the oil comes into contact with the skin, eugenol acts as a 'quencher' for the irritation reaction that some people have towards cinnamaldehyde. Indeed I.F.R.A. (International Fragrance Research Association) in their published guidelines on perfumery, recommend the addition of eugenol, glycol or phenyl ethyl alcohol at 1:1 ratio with cinnamon bark oil to desensitise the potential irritation.

It should be borne in mind that the leaf oil is one twentieth the price of the bark oil from cinnamon. Any prospective purchaser should be aware of this fact, making a selection of oil not only based on quality but value of natural cinnamaldehyde.

3.2.3 *Storage*

Plant materials, once prepared for storage and transportation by drying, curing, ageing, etc. should be kept under appropriate conditions. For instance, sacks of peppercorns which are stored under damp conditions, without free circulation of air, develop a 'musty' note which remains in the oil even after extraction. Some plant materials require several months of curing and maturation and are graded on the quality of this preparation (e.g. vanilla beans; details of their processing are discussed in Section 3.2.7).

3.2.4 *Yield*

Some plant materials, particularly seeds (e.g. coriander, celery, nutmeg) contain fixed oils (triglycerides of fatty acids) as well as essential oils. As extraction solvents have differing polarities and hence affinities for the components of plant materials other than essential oil, it is important to know the yield as well as the solvent. For example some commercial CO_2 extracts of celery seed are alcohol soluble while others are not. As alcohol solubility is often used as an analytical criterion for the assessment of essential oils, alcohol insolubility would classify such extracts as oleoresins [5, 6].

Table 3.2 Yield of extract from three species by different solvents, taken from [11]. It is regrettable that carbon dioxide solvent was not included in this excellent study

Solvent	Lovage root		Celery seed		Carrot seed	
	Concrete %	Oil %	Resinoid %	Oil %	Resinoid %	Oil %
1,1,2-Trichloro--1,2,2-trifluoroethane[a]	0.53	0.3	0.71	0.16	1.10	0.37
Methyl furan[b]	2.61	0.39	–	–	4.79	0.47
Ethanol	8.24	0.38	13.23	0.86	3.30	0.87
Dichloromethane	0.89	0.29	2.0	0.13	1.70	0.18
Hexane	1.15	0.34	–	–	–	–
Benzene[c]	1.03	0.39	–	–	–	–
Cyclohexane	1.70	0.24	–	–	–	–
Hydrodistillation	–	0.15	–	0.69	–	0.40
(without recovery of water soluble components)						

[a] Not a food grade solvent, must be considered as experimental only.
[b] Not a food grade solvent.
[c] Not a food grade solvent, toxic residues.

The yield can vary significantly with the solvent employed. For instance, the acetone extract of ginger root (*Zingiber officinale*) yields 7% where as CO_2 yields only 3%. CO_2 extracts all of the essential oil in an undegraded form, together with the gingerols and shogaols responsible for the pungency [7–9] because they have a molecular weight below 400, the cut off for liquid CO_2 extraction. Solutions of CO_2 extracted ginger are water soluble. Acetone, being more polar than CO_2, will extract resins which increase the weight of extract but not its organoleptic quality. However, for application to baked goods or cooked products, the extra resins of the solvent extract act as a fixative and help prevent other components from volatilising during the manufacture of the food product, so are desirable [10]. Solutions of acetone extracted ginger are *not* water soluble.

Cu [11] illustrates how much the yield and composition of lovage root, celery seed and carrot seed can vary with solvent polarity (Tables 3.2, 3.3).

3.2.5 *Degradation*

Some methods of preparing extracts cause degradation or changes of components. The time honoured method of steam distillation causes many such degradations, such as formation of terpineol isomers, p-cymene from gamma-terpinene, loss of citral, loss of esters, etc. [16]. This is because the conditions of wet steam and elevated temperatures that are employed, cause bond rearrangements. It should not be forgotten that even ethanol causes changes, especially if free acids are present in the plant material and esterification can take place [17]. Such changes are even considered desirable for some applications such as alcoholic beverages [18].

Formation of acetals, Schiff's bases, Strecker degradation compounds, Maillard reactions and other artefacts can also take place during processing, but not necessarily to the detriment of the final use of the extract in food. Many cooked food flavours are dependent on these types of reaction taking place during cooking. However, for true fresh natural tastes, processing without heating is more desirable.

3.2.6 Preparation of Plant Material

Prior to extraction, many preparation techniques are used to optimise the flavour of plant material.

3.2.6.1 *Drying.* This removes the bulk of water, enabling a larger batch of material to be processed. Drying also minimises mould spoilage, e.g. fresh culinary green ginger rhizomes are considerably more susceptible to moulds than dried root.

3.2.6.2 *Grinding.* This reduces particle size to increase surface area and allows better penetration of solvent. Care should be exercised to ensure that finely divided material is not left unprotected so that evaporation and oxidation take place. Gaseous carbon dioxide is most suitable for protection or 'blanketing' of ground plant materials.

3.2.6.3 *Chopping.* Some plants are better chopped than ground before extraction. This is the case when tannins or other astringent components are present that are not required in an extract intended for use in, for example, dairy products.

3.2.6.4 *Roasting.* Examples of pre-roasting before flavour extraction are coffee and kola nuts. It should be noted that decaffeinated coffee is roasted *after* the removal of caffeine from the green beans with supercritical CO_2. This ensures that the full roast flavour is present in the coffee even though the caffeine has been removed.

3.2.6.5 *Fermentation.* Often fermentation of plant material takes place naturally during drying, e.g. grass to hay, plums to prunes, etc. and is an essential process for some foods to exist at all, e.g. black tea from green tea leaves.

3.2.6.6 *Curing.* Forms include salting of fish (Scandanavian gravlax salmon), hot and cold smoking of meats, fish, cheese and preservation of meats. Some curing is only a controlled fermentation, e.g. vanilla beans.

Table 3.3 Comparison of extraction yields by % different techniques

Botanical	Part	Source	Origin	Steam dist	Liq CO_2*	Solvent (specified)
Ambrette	Seed	Hibiscus abelmoschus	Africa	0.2–0.6	1.5	—
Angelica	Seed	Angelica archangelica	N. Europe	0.3–0.8	3	—
Aniseed	Seed	Anisum pimpanellum	N. Europe	2.1–2.8	7[a]	15 (ethanol)
Anise star	Seed	Illicium verum	China	8–9	10	28 (ethanol)
Basil	Leaf	Ocimum basilicum	N. Europe	0.3–0.8	1.3[b]	1.5 (ethanol)
Caraway	Seed	Carum carvi	N. Europe	3–6	5	20 (ethanol)
Carob	Fruit	Ceratonia siliqua	Africa	<0.01	0.1	40 (water–ethanol)
Cardamom	Seed	Elletaria cardamomum	Guatemala	4–6	4 (5.8[b], 9.4[c])	10 (ethanol)
Cassia	Bark	Cinnamomum cassia	China	0.2–0.4	0.6	5 (dichloromethane)
Celery	Seed	Apium graveolens	India	2.5–3.0	3	13 (ethanol)
Cinnamon	Bark	Cinnamomum zeylanicum	Sri Lanka	0.5–0.8	1.4[a]	4 (dichloromethane)
Clove bud	Flower	Eugenia caryophyllata	Madagascar	15–17	16 (20[c])	20 (ethanol)
Chilli	Fruit	Capsicum annum	India	—	4.9[c]	10 (acetone) 30 (60% ethanol)
Coffee	Seed	Coffea arabica	Arabia	<0.01	5	—
Carrot	Seed	Daucas carrota	European	0.2–0.5	—	3.3 (ethanol)
Cocoa (defatted)	Seed	Theobroma cacao	Africa	<0.01	0.5	1 (ethanol)
Coriander	Seed	Coriandrum sativum	Romania	0.5–1.0	3 (1.3[b])	20 (ethanol)
Cumin	Seed	Cuminium cyminum	India	2.3–3.6	4.5[c]	12 (ethanol)
Cubeb	Fruit	Piper cubeba	E. Indies	10–16	13	12 (ethanol)
Fennel	Seed	Foeniculum dulce	Europe	2.5–3.5	5.8[c]	15 (ethanol)
Fenugreek	Seed	Trigonella foenum graecum	India	<0.01	2	8 (ethanol)

Ginger	Root	Zingiber officinalis	Jamaica	1.5–3.0	3 (3.7[c])	6.5	(ethanol)
Ginger	Root	Zingiber officinalis	Nigeria	1.5–3.0	3 (4.6[a])	7	(acetone)
Hop	Fruit	Humulus lupulus	England	0.3–0.5	12	20	(ethanol)
Lovage	Root	Levisticum officinale	European	0.1–0.2	—	8	(ethanol)
Juniper	Berry	Juniperus communis	Yugoslavia	0.7–1.6	2.7 (7.2[b])	—	
Marjoram	Leaf	Majorlana hortensis	N. Europe	0.2–2.0	1.7[b]	—	
Mace	Aril	Myristica fragrans	W. Indies	4–15	16 (18.5[c])	40	(expression)
Nutmeg	Seed	Myristica fragrans	W. Indies	7–16	16	45	(expression)
Oregano	Leaf	Origanum vulgare	European	3–4	5[b]	—	
Pepper	Fruit	Piper nigrum	India	1.0–2.6	4	18	(acetone)
Pepper	Fruit	Piper nigrum	Malaysia	1.0–2.0	3.5	10	(acetone)
Pimento	Berry	Pimenta officinalis	Jamaica	3.3–4.5	4.5 (5.3[b])	6	(ethanol)
Parsley	Seed	Petroselinum crispum	India	2.0–3.5	9.8[c]	20	(ethanol)
Poppy	Seed	Papaver somniferum	India	—	2.5[c]	50	(ethanol)
Rosemary	Leaf	Rosmarinus officinalis	Europe	0.5–1.5	7.5[b]	5	(methanol)
Sage	Leaf	Salvia officinalis	Europe	0.5–1.1	4.3[b]	8	(ethanol)
Savoury	Leaf	Satureja hortensis	Europe	1–2	0.8[b]	—	
Thyme	Leaf	Thymus vulgaris	Europe	1–2	2.1[b]	10	(ethanol)
Tea	Leaf	Thea sinensis	Africa	<0.01	0.2	35	(ethanol/water)
Turmeric	Root	Curcuma longa	India	5–6	3.4[c]		(ethanol)
Vanilla	Fruit	Vanilla fragrans	Madagascar	<0.01	4.5	25–45	(ethanol/water)
Vetiver	Root	Vetiveria zizanoides	W. Indies	0.5–1	1.0 (1.0[c])	—	(ethanol/water)

*CO_2 yields are own results except: [a] supercritical CO_2 reported yield [13]; [b] supercritical CO_2 + entrainer yield [13]; [c] subcritical CO_2 [14,15].

3.2.7   Vanilla Bean Curing (Classic Method) [19]

After the harvest the vanilla beans have no aroma. This develops only during the course of the curing, which should begin immediately upon picking in order to prevent the beans from bursting (splitting). The preparation consists of three processing steps.

3.2.7.1   Scalding.

Batches of 25–30 kg beans are set into plaited wire baskets and dipped into hot water at 60°C for 2 min. This not only interrupts the maturing process in the cellular tissue but also distributes the sap and enzymes, which are important for the formation of aroma.

3.2.7.2   Fermentation process.

The drained, still hot beans are laid in cases lined with wool blankets, in which they perspire for 24 h and lose part of their excess moisture. With the fermentation process, the forming of aroma begins and the beans now turn a chocolate brown colour.

3.2.7.3   Drying and conservation.

On the following day, the beans are dried on trays in the sun from 10 am until 3 pm. Then the trays are stacked in storerooms until the next morning in order to maintain a consistent temperature for the efficient fermentation by enzymes for as long as possible. This procedure is repeated, depending on the location and weather, for up to 8 days. Then the beans are dried on racks in a bright and airy storeroom for up to 3 months, before they are packed in wooden trunks to season and develop their aroma for another 2–3 months. The main task of the preparer is the constant checking of the beans to determine the right time for these operations and to notice possible mould formation at a very early stage, removing any mouldy beans quickly.

As the classical curing method is strongly dependent on the weather conditions in Madagascar, an industrial method has been developed which is occasionally used. Figure 3.1 describes a building for the preparation of 180 tons of green vanilla, equivalent to 40 tons of prepared vanilla, for use during the two-month harvesting season.

(a) Arrival of green vanilla.
(b) Storage of green vanilla. The two pre-storage rooms each take 3 tons of beans piled 1 m high for a period of 48 h (to avoid uncontrolled fermentation).
(c) Sorting. With the help of a conveyer belt (5 m long), ten workers sort the beans into the following grades: 1, burst (split beans); 2, beans shorter than 12 cm; 3, unripened beans; 4, beans of 2nd choice; 5, beans of 1st choice.
(d) Basketing. After sorting, the beans are put into cylindrical wire baskets each holding up to 50 kg.

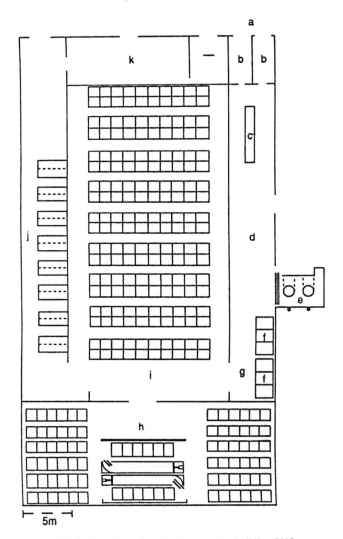

Figure 3.1 Plan of vanilla preparation building [20].

(e) scalding. The wire baskets are dipped into 63–65°C hot water, heated by a wood fire.

(f) Perspiring cases. After draining, the baskets are emptied into cases lined with wool blankets. The beans must be carefully and evenly piled in order to maintain a uniform temperature of about 50°C within the hermatically closed cases. This procedure kills the growth substances in the bean.

(g) Evaporating. After 48 h the beans are spread out on plaited shelves which are spaced vertically at 6 cm distances in a mobile rack so that air may circulate between the shelves. Every rack contains 22 shelves each of 11–12 kg vanilla beans.

(h) Drying tunnels (electrically heated to 65°C). Both drying tunnels take six mobile racks. Every 30 min a new rack is driven into the tunnel, so that the passage of each rack takes 3 h. For the period of 8 days every rack is passed each day through the drying tunnel.

(i) Drying room. The beans are now oily and ductile and are further dried depending on their moisture content. They are stored for 2 months on shelves of mobile racks, in a shady and well ventilated drying room.

3.3 Solvents

The choice of solvents available for food use is becoming more and more restricted. Recent recommendations of the EEC Solvents Directive have suggested that only the following solvents are used and they give advice on levels of residual solvents which are 'tolerable' in finished foods [21] (Tables 3.4, 3.5).

3.3.1 *Polarity*

A useful guide to the profile of components extractable by a solvent is its polarity. Armed with the knowledge of the constituents of a plant, it is possible to predict which components will be extracted under a given set of extraction conditions.

Table 3.4 Solvents and allowable residues in foodstuffs

Solvent	Residues (ppm)
Ethanol	Technically unavoidable quantities
Butane, propane	Technically unavoidable quantities
Acetone	Technically unavoidable quantities
Butyl and ethyl acetates	Technically unavoidable quantities
Nitrous oxide	Technically unavoidable quantities
Carbon dioxide	Technically unavoidable quantities
Dichloromethane	0.1 ppm (1 ppm confectionery/pastry 5 ppm decaffeinated tea, 10 ppm decaffeinated coffee)
Hexane	1 ppm (5 ppm oil/cocoa, 10 ppm defat flour, 30 ppm defat soya products)
Methanol, propan-1-ol, propan-2-ol	Levels to be reviewed June 1991
Methyl acetate	1 ppm (20 ppm in decaffeinated tea and coffee)
Ethyl methyl ketone	1 ppm (5 ppm fat or oil, 20 ppm decaffeinated tea and coffee)
Isobutane, cyclohexane, Butanols	1 ppm
Diethyl ether	2 ppm

Table 3.5 Solvent comparison [23]

	Viscosity (cP)		Latent heat evap (cal/g)	Boiling Point (°C)	Polarity (E°C)
	0°C	20°C			
Carbon dioxide	0.10	0.07	42.4	−56.6	0
Acetone	0.40	0.33	125.3	56.2	0.47
Benzene[a]	0.91	0.65	94.3	80.1	0.32
Ethanol	1.77	1.20	204.3	78.3	0.68
Ethyl acetate	0.58	0.46	94	77.1	0.38
Hexane	0.40	0.33	82	68.7	0
Methanol	0.82	0.60	262.8	64.8	0.73
Dichloromethane	–	0.43	78.7	40.8	0.32
Pentane	0.29	0.24	84	36.2	0
Propan-2-ol (IPA)	–	2.43	167	82.3	0.63
Toluene[a]	0.77	0.59	86	110.6	0.29
Water	1.80	1.00	540	100.0	>0.73
Propylene glycol	–	56.00	170	187.4	>0.73
Glycerol	12110.0	1490.0	239	290.0	>0.73

[a] Not food grade solvents.

3.3.2 Boiling Point

The temperature at which a solvent will boil under a given pressure or vacuum system, will influence the profile of an extract. For instance the change from methanol to ethanol to isopropanol to butanol at atmospheric pressure will alter the profile of molecular weight extracted. The higher the boiling point of reflux conditions, the higher the molecular weight profile that will be extracted.

3.3.3 Viscosity

The ability that a solvent has to penetrate plant material will depend on its mobility, hence viscosity. This property enables efficient extraction because of lower residence time of contact between solvent and plant material. The lower the viscosity of the solvent, the more efficient the commercial extraction.

3.3.4 Latent Heat of Evaporation

Costs of extraction are often limited to energy costs, and solvents of low latent heat of evaporation have the advantage of being removed using less energy than solvents of higher latent heats.

3.3.5 Temperature/Pressure

Besides being linked to boiling points, the temperature of reflux of a solvent in a given system is also dependent on pressure. As a guide, the temperature of a distillation will drop by 15°C every time the pressure is halved, so it is possible to accomplish some distillations at room temperature if the vacuum is high enough. This property is utilised for removal of residual solvent in the 'falling film evaporator'. This technique is an application of 'molecular distillation' or short path distillation [22].

3.4 Tinctures

3.4.1 *Water Infusions*

The use of water to extract or 'infuse' plant material such as roasted coffee and fermented tea leaves has been known since antiquity. Infusions of cocoa beans and vanilla beans were treasured by the Aztec Indians of Mexico and indeed were served to the conquistadors of Spain as a mark of respect. They

Figure 3.2 Solvent extraction plant.

Table 3.6 Water infusions

Examples	Source	Applications
Cola nuts	*Cola acuminata*	Soft drinks, cola drinks
Chrysanthamum flowers	*Chrysanthamum* sp.	Chrysanthamum tea drink
Coffee	*Coffea robusta, Arabica*	Soft drinks, candy, liqueurs
Tea	*Thea sinensis*	Soft drinks
Cocoa	*Theobroma cacao*	Soft drinks, liqueurs
Licorice root	*Glycyrrhiza glabra*	Candy, pharmaceutical products
Carob bean	*Ceratonia siliqua*	Chocolate flavourings, baked goods, tobacco flavourings, nut and rum flavourings
Rhubarb	*Rheum officinale*	Soft drinks, pharmaceutical products
Elderberries	*Sambucus nigra*	Red fruit flavourings

introduced the beverage to Europe and the rest of the World.

The technique for the preparation of infusions is simple, using either hot or cold water which remains in the finished extract. The plant material is ground to a broad spectrum particle size to facilitate the flow of water through the bed of material. This is important because if the particles are all too fine, the bed will compact and not allow the passage of solvent. In the specific case of infusions, there is often some swelling of the material when it comes into contact with water which slows the flow of solvent. This softening is known as 'maceration'. The material is allowed to stand in a 'percolator' (see Figure 3.2) for approximately 1 week. The liquid 'menstruum' is drawn off from a tap which is located at a position low on the side of the percolator. Fresh solvent is added to the percolator to give successive 'coverings'. All the menstruums of varying strengths are then combined and standardised for flavour and non-volatile residue. This is the completed infusion. Some examples are listed in Table 3.6.

3.4.2 *Alcoholic Tinctures*

By a process analagous to infusions but using ethanol, propan-2-ol, propylene glycol or glycerol as a solvent, a tincture is prepared. Unlike an oleoresin, the solvent is left as part of the product which is then incorporated directly into a food, beverage or pharmaceutical product.

An adaptation of the 'maceration' technique is sometimes used where the chopped plant material is placed in a gauze bag and the solvent is pumped through it. This 'circulatory maceration' is usually more rapid than static percolation and offers process advantages with some materials e.g. vanilla beans and other expensive raw materials.

Many such tinctures of herbs and spices are used in the soft drinks and

Table 3.7 Alcoholic tinctures

Examples	Source	Applications
Benzoin resin	*Styrax benzoin*	Tobacco flavours, chocolate and vanilla flavours
Quassia bark	*Picrasma excelsa*	Bittering agent for soft drinks and alcoholic beverages
Buchu leaves	*Barosma betulina*	Blackcurrant flavours
Strawberry leaves	*Fragaria vesca*	'Green' strawberry flavours
Blackcurrant leaves	*Ribes nigrum*	Blackcurrant flavours
Elderflowers	*Sambucus nigra*	Beverages
Banana fruit	*Musa* sp.	Tropical fruit flavours
Ginger root	*Zingiber officinale*	Beverages
Orris root	*Iris germanica*	Tobacco flavours, soft drinks

alcoholic beverage industries [24] to impart mouth feel and taste, as well as the flavour that could have been achieved with a solution of the essential oil. This list is extensive, from the exotic southernwood, dittany of Crete and chrysanthemum flowers to the more mundane ginger, kola and quassia bark products (Table 3.7).

3.5 Oleoresins

Some oleoresins occur naturally as gums or exudates directly from trees and plants. Examples are benzoin (*Styrax benzoin*), myrrh (*Commiphora molmol*), copaiba balsam (*Copaefera*), olibanum (*Boswellia carteri*) and other balsams. They are usually only partially solvent soluble and are usually extracted before use, when they are known as 'prepared oleoresins'.

3.5.1 Solvents

Those commonly employed have already been listed, i.e. ethanol, water (and mixtures), isopropanol, methanol, ethyl acetate, acetone. The preparation of oleoresins closely follows the methodology for tinctures and infusions.

Oleoresins, resins and concretes have their extraction solvent removed with only minimal residues remaining. Attempts to remove these last traces of solvents usually result in the loss of some of the 'top notes' from the oleoresin. This is not important if the primary use of the oleoresin is for pungency, e.g. capsicum (*Capsicum annuum*). For most other applications, such as pepper and ginger oleoresins, it is preferable to tolerate solvent residues rather than lose top notes from the oil fraction that make a valuable contribution to flavour in a food application. Concretes differ from oleoresins only in that they are more crystalline in physical form and are made by the same technique. An exception is orris concrete (*Iris germanica*) which is distilled but contains myristic acid (tetradecanoic acid) and sets to a soft crystalline mass at ambient temperature.

3.5.2 Solubility

Depending on the boiling point and polarity of the solvent, a balance between the molecular weight and polarity of the component profile of the plant material will be extracted with time. For example with a hydrocarbon solvent, the low molecular weight (MW), low polarity components extract first, and presuming that sufficient solvent is available so that saturation does not occur, medium MW, low polarity components and low MW, polar components then extract. The successively high MW, low polarity and medium MW, polar components etc. are extracted until saturation is reached. In practice, sufficient solvent is circulated from the start of the extraction to remove all desirable components and this fractionation effect is not normally observed.

3.5.3 Commercial Solvent Extraction System

A commercial solvent extraction system is shown in Figure 3.2. The oleoresins listed in Table 3.8 are typical examples of the many produced commercially, but there is considerable overlap between the drug, pharmaceutical, food and beverage industries.

3.5.3.1 *Standardised oleoresins* (Figure 3.3). Some commercial oleoresins are not what they seem. The high level of lipids in some plants means that a total extract would be very weak in flavour. Such extracts are standardised by addition of steam distilled oil to reinforce the 'flavour' as well as providing the 'taste' of the spice. Examples of these 'standardised oleoresins' are coriander, bay, nutmeg, mace, pepper. They are also available in the form of emulsions, spray dried encapsulated or simply 'spread' or 'plated' onto salt, rusk or

Table 3.8 Commercially produced oleoresins

Examples	Source	Applications
Pepper oleoresin	*Piper nigrum*	Snacks, curries, sauces, pickles
Pimento oleoresin	*Pimenta officinalis*	Pickles, chutneys, sauces, spice blends
Tonka concrete	*Dipteryx odorata*	Tobacco flavours (restricted)
Vanilla oleoresin	*Vanilla fragrans*	Liqueurs, natural baked goods, sweet sauces
Ginger oleoresin	*Zingiber officinale*	Baked goods
Fenugreek oleoresin	*Trigenella* f.g.	Spice blends, curries
Lovage oleoresin	*Levisticum officinale*	Soups, savory sauces
Nutmeg oleoresin	*Myristica fragrans*	Soups, savory and sweet sauces
Turmeric oleoresin	*Curcuma longa*	Spice blends, natural colour
Oregano oleoresin	*Origanum vulgare*	Pizza toppings, herb sauces
Rosemary oleoresin	*Rosemarinus officinalis*	Natural antioxidant, meat sauces

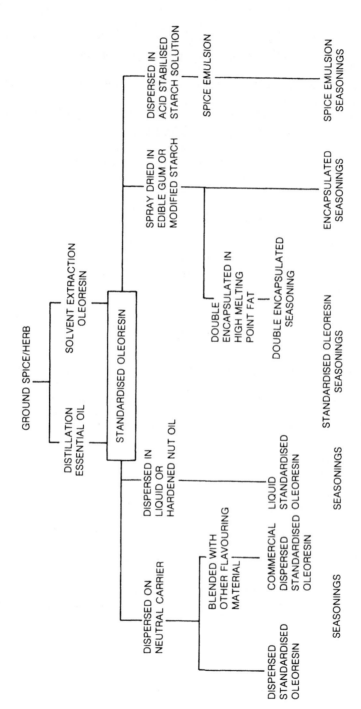

Figure 3.3 Flow chart of a typical standardised oleoresin range of spices.

Table 3.9 Standardised oleoresins[a]

		Volatile oil content (%, v/w)	Other standards where applicable
Basil	FD0739	9–12	
Bay	FD3490	32–36	
Capsicum BPC for pharmaceutical application[b]			Capsaicin 9.0–11.0%
Capsicum BPC 1923		–	Capsaicin min 1.25%
Capsicum	FD5388	–	Capsaicin 5.0–6.0%
Capsicum	FD5850	–	Capsaicin 2.5–3.0%
Cayenne	FD3576	–	Capsaicin 12.75–14.0%
Capsicum red	FD3295	–	Capsaicin 2.4–3.0% Colour value min: 0.26 = 15,800 c.u.
Capsicum	FD5835		Capsaicin 9.5–10.5%
Caraway	FD5458	33–37	
Cardamom	FD3293	52–58	
Celery	FD3299	12–14	
Celery	FD3774	40–45	
Chervil	FD1189	N/A[c]	
Cassia concentrate	EF9509	5.7–7.2 powder	
Cinnamon concentrate	FD6258	5.7–7.2 powder	
Clove USF	FD3405	74–80	
Coriander	FD3292	38–42	
Coriander	FD0743	50–56	
Cumin	FD3406	63–70	
Cumin	FD0744	29–34	
Dill	FD0745	43–47	
Dill weed	FD3775	30–34	
Fennel	FD1250·	7.5–10	
Fenugreek	FD2959	N/A	
Garlic liquid concentrate	FD6192	N/A	
Ginger African	FD3297	Minimum 25	
Ginger Jamaican	FD3296	Minimum 25	
Ginger	FD3298	Minimum 25	
Gingerine water Soluble		N/A	
Gingerine select		Minimum 25	
Lovage	FD1220	10–20	
Mace	FD5407	34–42	

Table 3.9 (*cont'd over*)

Table 3.9 (*Contd.*)

		Volatile oil content (%, v/w)	Other standards where applicable
Mace	FD5847	66–72	
Mace	FD0721	26–30	
Mace	FD5698	50–54	
Marjoram	FD0761	10–12	
Mushroom	FD5897	N/A	
Nutmeg	FD5678	90–95	
Nutmeg	FD3264	80–90	
Nutmeg	FD5825	85–95	
Onion	FD6039	N/A	
Onion	FD5323	N/A	
Oregano	FD3777	36–44	
Origanum	FD1451	18–23	
Paprika	H8026		Colour value: min 0.455 = 27,750 c.u.
Paprika	FD3912		Colour value: min 0.656 = 40,000 c.u.
Paprika	FD0722		Colour value: 0.3–0.35 = 20,000 c.u.
Parsley	FD4098	7.5–9	
Pepper liquid	FD5044	14.3–15.8	Piperine 40.4–44.0%
Pepper pourable	FD5640	21–23	Piperine 40.0–44.0%
Pepper pourable	FD5662	21–23	Piperine 40.0–44.0%
Pimento	FD3262	60–66	
Pimento	FD0754	46–52	
Rosemary	FD3493	34–40	
Sage	FD3493	22–27	
Sage Dalmatian	FD0750	38–43	
Savory	FD0867	7–9	
Spice mixed	FD5232	N/A	
Spice mixed	FD5613	N/A	
Tarragon	FD3484	14–18	
Thyme	FD3484	54–60	
Thyme	FD0718	15–20	
Turmeric water soluble			Curcumin 5.0–7.0
Turmeric	FD6454		Curcumin approx. 30%
Turmeric	FD3778		Curcumin 38.0–42.0%
Turmeric	FD3487		Curcumin 12.5–14.5%

[a] A leaflet entitled *Standardised Oleoresins* detailing the advantages, application, outline specifications, and analytical methods is available upon request from Bush Boake Allen Limited, London, England.
[b] Pungency of capsicum products: 1% capsaicin is equivalent to 127,000 Scoville units.
[c] N/A, not applicable or no valid standard.

Table 3.10 Standardised oleoresins[a]

		Volatile oil content (%, v/w)	Other standards where applicable	Dispersion rate kgs = 100 kg of spice
Aniseed	HX2352	50		3
Basil	HX2114	40		1
Bay	HX2117	30		3
Caraway	HX2224	60		1
Cardamon	HX2119	50		4
Cassia bark	HX2093	50		3
Cayenne pepper	HX2110		Pungency 1 M s.u. Colour 500 c.u.	2.5
Celery seed	HX2120	7.5		4
Cinnamon bark	HX2116	45		2
Clove bud	HX2118	50		10
Coriander seed	HX2041	33		3
Cumin seed	HX2338	60		5
Ginger Nigerian	HX2084	30		3.5
Ginger Jamaican	HX2088	25		2.5
Ginger Cochin	HX2208	25		3.5
Mace West Indian	HX2132	25		10
Marjoram	HX2128	12		6
Nutmeg	HX2042	80		5
Oregano	HX2352	18		8
Paprika	HX2111		60,000 c.u.	4.5
Parsley	HX2091	2		3
Pepper	HX2461	20	40% piperine	7
Pimento	HX2112	60		3.5
Rosemary	HX2113	30		4
Sage Dalmatian	HX2090	65		2
Tarragon	HX2221	14		2
Thyme	HX2089	50		1

[a] A range of standardised oleoresins is available from Lionel Hitchen Essential Oil Company Limited, Barton Stacey, Hants, UK.

Table 3.11   Dispersed spices—salt

		Volatile oil content (% v/w)
Basil		0.25–0.35
Bay		1.0–1.2
Bouquet garni	FD1360	N/A
Caraway		1.5–1.8
Cardamom		2.8–3.2
Cassia		0.6–0.8
Cayenne	Capsaicin	0.28–0.32% w/w
Celery		1.0–1.3
Green celery	FD3309	0.3–0.4
Chervil		N/A
Cinnamon		0.6–0.8
Clove		9.0–10.0
Coriander		0.45–0.55
Cumin		1.0–1.3
Dill		1.4–1.6
Fennel		0.6–0.8
Fenugreek		N/A
Garlic		N/A
Ginger	FD528	0.9–1.2
Juniper	FD1193	1.3–1.5
Leek		N/A
Lovage		0.2–0.4
Mace	FD529	2.3–2.8
Marjoram		0.35–0.5
Mint		0.3–0.6
Mushroom	FD5898	N/A
Nutmeg		3.5–4.5
Onion		N/A
Oregano		1.4–1.7
Paprika	Colour value	0.30–0.35
Parsley		N/A
Green parsley		N/A-
Black pepper	FD513	V.O. 1.4–1.7
		Piperine 3.0–4.0% w/w
Pepper		V.O. 1.0–1.3
		Piperine 3.5–4.0% w/w
Pimento		1.6–2.0
Rosemary		0.7–0.9
Green rosemary	FD610	1.0–1.2
Sage Dalmatian		1.0–1.3
Savory		0.25–0.35
Tarragon		0.2–0.4
Thyme		0.3–0.4
Turmeric	Curcumin	0.6–0.7% w/w

Information on 'Saromex' and Bush Boake Allen's spice extractives publication, *Specialisation in Spices* is available in several languages.

Table 3.12 Dispersed spices—dextrose

		Volatile oil content (%, v/w)
Bay		1.0–1.2
Cayenne	Capsaicin	0.28–0.32% w/w
Celery		1.0–1.3
Cinnamon		0.6–0.8
Clove		9.0–10.0
Coriander		0.45–0.55
Cumin		1.0–1.3
Garlic		N/A
Ginger	FD5732	0.8–1.1
Mace		1.9–2.1
Marjoram		0.35–0.50
Mixed spice	FD2423	N/A
Nutmeg		3.5–4.5
Pepper		V.O. 1.0–1.3
		Piperine 3.5–4.0% w/w
Thyme		0.3–0.4

Table 3.13 Dispersed spices—rusk

		Volatile oil content (%, v/w)
Bay	FD5771	0.8–1.0
Cayenne	Capsaicin	0.28–0.32% w/w
Celery		1.0–1.3
Clove		9.0–10.0
Coriander		0.45–0.55
Ginger	FD541	1.3–1.5
Mace		3.8–4.2
Marjoram		0.35–0.50
Nutmeg		3.5–4.5
Pepper		V.O. 1.0–1.3
		Piperine 3.5–4.0% w/w
Pimento		1.6–2.0
Thyme	FD5781	0.6–0.8

Table 3.14 Standardised emulsion oleoresins[a]

			Strength compared to ground spice
Garlic	Ail	HF101	10 ×
Aniseed	Aneth	HF145	4 ×
Basil	Basilic	HF128	4 ×
Cinnamon	Cannelle	HF114	4 ×
Cardamom	Cardamone	HF129	4 ×
Caraway	Carvi	HF150	4 ×
Cayenne	Cayenne	HF122	4 ×
Celery	Celeri	HF112	4 ×
Chervil	Cerfeuil	HF146	2 ×
Mushroom	Champignon	HF179	0.3 in hydrolysed vegetable protein
Mushroom	Ciboulette	HF149	4 ×
Coriander seed	Coriandre	HF103	4 ×
Coriander leaf	C. Feuilles	HF432	4 ×
Cumin	Cumin	HF123	4 ×
Curry	Curry	HF198	5–6 g/kg
Shallot	Eschalote	HF118	10 × (HF 144 4 ×)
Tarragon	Estrayon	HF115	10 × (HF 170 4 ×)
Fennel	Fenouil	HF245	2 ×
Smoke	Fume	HF148	0.5/1 g/kg
Juniper	Genievre	HF185	4 ×
Ginger	Gingembre	HF124	4 ×
Clove bud	Girofle	HF116	2 ×
Bay laurel	Laurier	HF108	4 ×
Mace	Macis	HF158	2 ×
Marjoram	Marjolaine	HF147	4 ×
Mint	Menthe	HF197	4 ×
Nutmeg	Muscade	HF111	4 ×
Onion	Oignon	HF138	10 ×
Origano	Origan	HF176	4 ×
Parsley	Persil	HF126	4 ×
Pimento	Pimento	HF125	4 ×
	Poireau	HF275	4 ×
Pepper black	Poivre	HF102	4 ×
Pepper green	P. Vert	HF276	2 ×
Rosemary	Roumarin	HF127	4 ×
	Sarriette	HF186	4 ×
Sage	Sauge	HF131	4 ×
	Serpolet	HF320	4 ×
Thyme	Thym	HF107	4 ×

[a] A range of liquid dispersed standardised and emulsified oleoresins, complete with specifications and applications data, is available from Felton Worldwide SARL, Versailles, France.

Table 3.15 Standardised emulsion oleoresins[a]

		Strength compared to equivalent ground spice
Basil sweet	FD6113	5 ×
Bay	FD6114	5 ×
Caraway	FD6115	5 ×
Cardamom	FD6116	5 ×
Cassia	FD6117	5 ×
Cayenne	FD6118	5 ×
Celery	FD6119	5 ×
Cinnamon	FD6120	5 ×
Clove	FD6121	2 ×
Coriander	FD6122	5 ×
Cumin	FD6123	5 ×
Dill	FD6153	5 ×
Fennel	FD6224	5 ×
Fenugreek	FD6315	5 ×
Garlic	FD6124	5 ×
Garlic	FD6227	5 ×
Ginger	FD6125	5 ×
Mace	FD6126	5 ×
Marjoram	FD6127	5 ×
Nutmeg	FD6128	5 ×
Onion	FD6284	50 ×
Oregano	FD6162	5 ×
Paprika	FD6129	3.5 × colour
Parsley	V0235	5 ×
Pepper	FD6336	3 ×
Pepper black	FD6220	3.5 ×
Pepper	FD6326	3 ×
Pepper	FD6206	4 ×
Pimento	FD6131	5 ×
Rosemary	FD6132	5 ×
Sage Dalmatian	FD6133	5 ×
Savory	FD6134	5 ×
Tarragon	FD6135	5 ×
Thyme	FD6136	5 ×
Turmeric	FD6137	2.5 × colour

[a] Bush Boake Allen also produce a liquid range.

Table 3.16 Encapsulated standardised oleoresin

		Approximate strength compared to spice[a]
Aniseed	FD6353	5 ×
Basil	FD4031	10 ×
Bay	FD3883	10 ×
Caraway	FD4032	10 ×
Cardamom	FD4033	10 ×
Cayenne	FD4051	10 ×
Celery	FD6349	10 ×
Green celery	FD4034	10 × (FD3309)
Cinnamon	V0337	10 ×
Clove	FD5013	5 ×
Coriander	FD4035	10 ×
Cumin	FD4047	10 ×
Dill	FD3885	10 ×
Fennel	FD4037	10 ×
Fenugreek	FD4036	10 ×
Garlic	FD5580	10 ×
Ginger	FD4038	10 × (FD528)
Leek	FD5789	50 ×
Mace	FD4054	10 × (FD529)
Marjoram	FD3941	10 ×
Mushroom	FD6351	5 × (FD5898)
Nutmeg	FD3335	10 ×
Onion	FD5450	10 ×
Oregano	FD6350	10 ×
Paprika	FD5704	10 ×
Parsley	FD5258	10 × (Green)
Pepper	FD5457	4 × (FD 513)
Pimento	FD4042	10 ×
Rosemary	FD4043	10 × (FD610)
Sage Dalmatian	FD4039	10 ×
Savory	FD4044	10 ×
Tarragon	FD3940	10 ×
Thyme	FD4040	10 ×
Turmeric	FD6141	10 ×

[a] These concentrated, salt-free, hygienic powders effectively 'lock-in' the flavour in dry mixes until the product is reconstituted: only when the encapsulant holding the flavour is wetted does the flavour become available. Areas of application include frozen/chilled foods, biscuits, snack foods, flour confectionery, cake mixes and in some instances, extruded products. Information is contained in the BBA entitled, 'Saronseal Encapsulated Spices'.

maltodextrin. It should be remembered that because of the blending with steam distilled oils, these 'standardised spices', although functional, are not a true representation of the plant material. Several manufacturers offer ranges of such standardised spices (Tables 3.9–3.16).

3.6 Absolutes

These are alcohol soluble extracts of plant materials which represent the 'heart' of the flavouring. Some examples are listed in Table 3.17.

3.6.1 Solvents

Absolutes can be prepared by one of the following.

(a) Extraction of the plant material with ethanol and evaporating the solvent.
(b) Extracting a crude whole oleoresin with a suitable solvent, evaporating the solvent and re-extracting the absolute from the oleoresin with alcohol. This has the process advantage of utilising readily available resins of the base and so minimising the use of alcohol.
(c) Some naturals can be extracted with CO_2 to give extracts with all of the attributes of an absolute but processed without organic solvents (see Section 3.7).

Table 3.17 Alcohol soluble extracts

Examples	Source	Applications
Rose absolute	*Rosa centifolia* (de Mai) *Rosa damascenia*	Confectionery, fantasy, soft drinks
Jasmin absolute	*Jasminum grandifolia*	Top note fruit flavours
Tobacco absolute	*Nicotiana* sp.	Tobacco flavours
Ginger absolute	*Zingiber officinalis*	Beverages, liqueurs
Spearmint absolute	*Mentha spicata*	Oral hygiene
Boronia absolute	*Boronia megastigma*	Fruit flavours
Deer tongue absolute	*Trilisa odoratissima*	Tobacco flavours
Vanilla absolute	*Vanilla fragrans*	Dairy, liqueurs
Orange blossom, absolute from waters	*Citrus aurantium*	Orange flavours modifier
Labdanum absolute	*Cistus ladaniferus*	Soft fruit flavours
Cassis absolute	*Ribes nigrum*	Natural blackcurrant flavours
Mimosa absolute	*Acacia decurrens*	Fruit flavours
Genet absolute	*Spartium junceum*	Fruit flavours

3.7 Extraction with Carbon Dioxide as a Solvent

3.7.1 Introduction

Fundamentally all dry botanicals with an oil or resin content can be extracted with carbon dioxide. This pressured solvent behaves during extraction in a similar way to any of the other solvents previously discussed. As a solvent it has some significant advantages compared to alternatives. It is:

odourless and tasteless
food safe and non-combustible
easily penetrates (because of its low viscosity)
easily removed with no solvent residue (because of its low latent heat of evaporation)
can be used selectively (by varying the temperature and pressure of extraction)
inexpensive and readily available

A comparison with traditional forms of extraction shows that carbon dioxide is versatile.

Traditional extraction

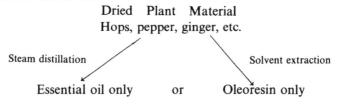

CO_2 extraction

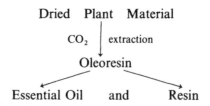

Commercially CO_2 can be used in two distinct modes of extraction which are dependent on its operation above or below the critical point in the phase diagram for CO_2 (Figure 3.4).

3.7.2 Subcritical CO_2

Using extraction conditions of 50–80 bar pressure and 0 to +10°C, it is commercially viable to extract essential oils as an alternative to steam

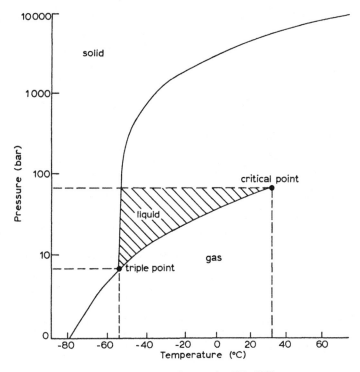

Figure 3.4 Phase diagram for CO_2 [25].

distillation. The energy savings of CO_2 explained in the section on solvents offset some of the capital expenditure of the extraction equipment.

3.7.2.1 *Solubility of components.* See Table 3.18.

Table 3.18 Solubility of botanical components in liquid CO_2. Liquid CO_2 extracts all of the useful aromatic components from botanical materials [25]

Very soluble	Sparingly soluble	Almost insoluble
Low MW aliphatic hydrocarbons, carbonyls, esters, ethers (e.g. cineole) alcohols monoterpenes, sesquiterpenes	Higher MW aliphatic hydrocarbons, esters, etc.; substituted terpenes and sesquiterpenes; carboxylic acids and polar N and SH compounds; saturated lipids up to C12.	Sugars, protein, polyphenols, waxes; inorganic salts; high MW compounds, e.g. chlorophyll, carotenoids, unsaturated and higher than C12 lipids
MW up to 250	MW up to 400	MW above 400

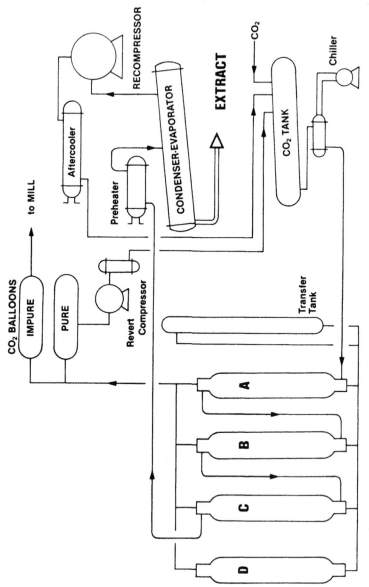

Figure 3.5 Extraction circuit CO_2.

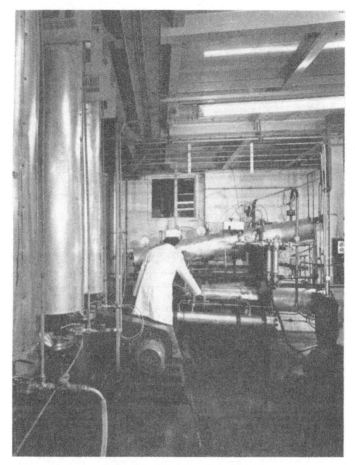

Figure 3.6 CO_2 extraction plant.

3.7.2.2 *Extraction circuit* [26].   See Figures 3.5, 3.6.

3.7.2.3 *Entraining solvents.*   'Entrainers' can be injected into the flow of CO_2 to increase its polarity for certain extractions, e.g. juniper berry oil [27]. This author considers that ethanol is the only suitable entrainer for food grade products, although others have been reported in the technical literature. The use of hexane and similar organic solvents, obviates a major advantage that CO_2 offers for foods, that of no solvent residues.

3.7.2.4 *Properties of extracts.*   These include: no solvent residues; no off notes; more top notes; more back notes; better solubility; concentration of aromatics.

Table 3.19 Applications of CO_2 extracts

Examples	Source	Applications
Ginger oil	*Zingiber officinalis*	Oriental cuisines, beverages
Pimento berry oil	*Pimenta officinalis*	Savory sauces, oral hygiene
Clove bud oil	*Eugenia caryophyllata*	Meats, pickles, oral hygiene
Nutmeg oil	*Myristica fragrans*	Soups, sauces, vegetable juices
Juniper berry oil	*Juniperus officinalis*	Alcoholic beverages, gin
Celery seed oil	*Apium graveolens*	Soups, vegetable juice (tomato)
Vanilla absolute	*Vanilla fragrans*	Cream liqueurs, pure dairy foods
Cardamom oil	*Elletaria cardamomum*	Meats, pickles, spice blends
Aniseed oil	*Illicium Verum*	Liqueurs, oral hygiene
Coriander oil	*Coriandrum sativum*	Curry, chocolate, fruit flavours

Table 3.20 Templar CO_2 extracts—main list[a]

		Origin	Plant Source
Cardamom	C1536	Guatemala	*Elletaria cardamomum*
Celery seed	C1919	India	*Apium graveolens*
Clove bud	C1671	Comores	*Eugenia caryophyllata*
Coriander seed	C1532	Roumania	*Coriandrum sativum*
Ginger Jamaican	C1537	Jamaica	*Zingiber officinale*
Ginger terpeneless	C1622	Nigeria	*Zingiber officinale*
Hop oil	C1567	England	*Humulus lupulus-target*
Hoparome absolute	C2009	England	*Humulus lupulus-target*
Juniper berry 20%	C1521	Yugoslavia	*Juniperus communis*
Nutmeg	C1676	West Indies	*Myristica fragrans*
Pepper oil black	C1522	Sarawak, Malaysia	*Piper nigrum*
Pimento berry	C1533	Jamaica	*Pimenta officinalis*
Vanilla absolute	C1577	Madagascar	*Vanilla fragrans*

[a]These low temperature CO_2 extracts are unblended, pure and 100% natural from the named origin. Full stock, technical specifications and data sheets are available with price lists, quoted as delivered prices in local currency. These main list items are available in bulk at competitive prices from Felton Worldwide Offices throughout the world.

3.7.2.5 *Applications.* See Tables 3.19–3.21.

3.7.3 *Supercritical CO_2*

Conditions are 200–300 bar pressure and + 50 to 80°C.

3.7.3.1 *Solubility of components.* As a generalisation, supercritical CO_2 can extract all the soluble components from plant material in a similar way to organic solvents, giving oleoresins. These oleoresins are free of organic solvent residues and can be further fractionated (see Section 3.7.3.3) to give oils. The

Table 3.21 Templar CO_2 extracts—supplementary list[a]

	Origin	Plant source
Ambrette seed	Africa	*Hibiscus abelmoschus*
Angelica seed	N. Europe	*Angelica archangelica*
Anise star	China	*Illicium verum*
Caraway	N. Europe	*Carum carvi*
Cubeb	East Indies	*Piper cubeba*
Coffee	Arabia	*Coffea arabica*
Cocoa	Africa	*Theobroma cacao*
Fenugreek	India	*Trigonella foenum graecum*
Ginger Australian	Australia	*Zingiber officinale*
Ginger WI	West Indies	*Zingiber officinale*
Ginger Jamaican terpeneless	Jamaica	*Zingiber officinale*
Ginger absolute	Nigeria	*Zingiber officinale*
Mace	West Indies	*Myristica fragrans*
Pepper extract, high aroma	Sarawak	*Piper nigrum*
Pepper extract, high piperine	Sri Lanka	*Piper nigrum*
Piperine crystals	Malaysia	*Piper nigrum*
Strawberry juice	N. Europe	*Fragaria vesca*
Tea	Africa	*Thea sinensis*

[a] These low temperature CO_2 extracts are unblended, pure and 100% natural from the named origin. Available as made to order or contract only. Lead times, availability and price quotation from Felton Worldwide Limited, Bletchley, UK.

profile of any 'supercritical CO_2 extract' should be studied carefully to ensure the desired characteristics are present.

3.7.3.2 *Costs.* These extracts, although desirably giving an oleoresin free of solvent residue, are at a considerable cost disadvantage to conventional solvent extracts. Daily processing costs are similar but the capital cost of a commercial scale, high pressure, supercritical plant is millions of pounds sterling, while solvent extraction equipment costs just thousands of pounds. Solvents oleoresins are more cost effective than supercritical extracts.

3.7.3.3 *Fractionations.* Supercritical CO_2 extraction does have one significant advantage over subcritical CO_2 or solvent extraction. This is the ability to fractionate the oleoresin by altering the pressure of the system and 'tapping' out the desired fraction. This concept was reported by Brogle [28] and applied in several experimental publications including a recent paper by Sankar [29] on pepper oil from supercritical CO_2 oleoresin.

3.7.3.4 *Applications.* Some applications and examples of supercritical CO_2 extracts are listed in Tables 3.22, 3.23.

Table 3.22 Supercritical CO_2 extracts

Examples	Source	Applications
Pepper oleoresin and oil [29]	*Piper nigrum*	Spices, meat, salad dressing
Clove bud oleoresin [30]	*Eugenia caryophyllata*	Oral hygiene, meats
Ginger [31, 32]	*Zingiber officinale*	Spices, sweet products
Cinnamon bark [13]	*Cinnamomun zeylendium*	Baked goods, sweet products
Lilac flower [13]	*Syringa vulgaris*	–
Cumin	*Cuminium cyminum*	Mexican and Indian cuisines
Marjoram	*Majorlana hortensis*	Soups, savoury sauces
Savory	*Satureja hortensis*	Soups, savoury sauces
Rosemary	*Rosemary officinalis*	Anti-oxidant, soups
Sage	*Salvia officinalis*	Meat, sauces, soups
Thyme	*Thymus vulgaris*	Meat, pharmaceutical products

Table 3.23 PFICO$_2$ Extracts. These are examples of CO_2 extracts available from CAL-Pfizer, Grasse, France, who indicate in their literature that they are prepared with supercritical CO_2

	Origin	Plant source
Celery seed	India	*Apium graveolens*
Ginger	India	*Zingiber officinale*
Ginger	China	*Zingiber officinale*
Paprika flavour	European	*Capsicum annum*
Paprika colour	European	*Capsicum annum*
Rosemary	France	*Rosmarinus officinalis*
Sage	France	*Salvia officinalis*
Vanilla absolute	Madagascar (or) Comores (or) Reunion	*Vanilla fragrans*

3.8 Summary

When selecting a natural product extract for inclusion into a flavour, the choice of extraction technique, based on a consideration of the end-product, will often prevent problems at a later stage of product development. The selection of solvent is often the key to obtaining an extract with the desired properties of flavour, taste and solubility required. The most natural, true tasting extracts of plant materials are obtained by the techniques which utilise low temperature and avoid degradative heat processes and reactive solvents.

4 Fruit juices

P.R. ASHURST

4.1 Introduction

There has been a long association between fruit juices and flavourings. Traditionally, fruit flavourings were some of the earliest types available and because of their relative simplicity, they have often been used to enhance or substitute for fruit juices in beverages.

Fruit juices and their components also play a very important part in many flavourings, with concentrated juices frequently used as a base to which other components may be added. With the growth in interest and demand for natural flavours, fruit juice components are an essential source of these ingredients, although they are rarely, if ever, combined in the same proportions as in the original fruit juice.

The biological function of fruits is to be attractive to animals to ensure distribution of the seed via animal faeces or, in the case of larger fruits, to provide a bed of rotting humus in which the seed may develop. In contrast to many other vegetable products such as cereals, the starting point for juice production is the tender fleshy fruit which is prone to more or less rapid decay. This instability is increased once the fruit is broken to initiate a process and in consequence, all man's early attempts to utilise fruit juices ended in fermented products such as wine or cider. Early in the nineteenth century, Appert (1725–1841) showed that fruit juices could be stabilised by heat treatment after bottling and in 1860 the discoveries of Pasteur provided a scientific background for this observation.

In both Europe and the United States, the commercial production of pasteurised fruit juices began late in the nineteenth century, but it was not until the second quarter of the twentieth century that technical and commercial development of fruit juices really began to be significant. With availability of fruit juices came increasing consumption and their incorporation into other products such as soft drinks. The world-wide availability of fruit juices is now taken for granted and the manufacturing industry is large, complex and well organised.

There has always been a close link between flavourings and fruit juices, with synthetic materials used to extend and enhance juices. Juices themselves, particularly concentrates and volatile fractions are being increasingly used as components of flavourings. This trend has become more noticeable as the demand for natural flavourings has increased.

4.2 Fruit Processing

4.2.1 *General Considerations*

It is necessary to review typical manufacturing processes for various types of fruit in some detail in order to become familiar with the different types of fruit raw materials that may be available. A typical fruit processing operation is shown in Figure 4.1.

Botanically, fruits are formed from floral parts of plants with or without associated structures. The tissues bearing juice are either thin walled parenchyma or special sacs such as the modified placental hairs of citrus. Fruit is often classified on the basis of the hardness of its pericarp (the wall of the

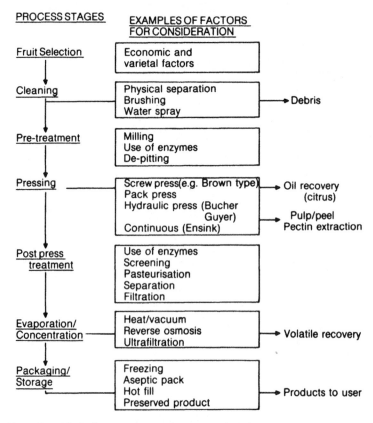

Figure 4.1 Block diagram of stages in a typical fruit juice processing operation.

ovary from which the fruit is derived). A typical classification is shown below (see also Appendix II):

Dry fruits   Dehiscent (bursting fruits), e.g. legumes
              Indehiscent (non-bursting fruits), e.g. nuts
Fleshy fruits   Berry, e.g. blackcurrant, citrus
              Drupe (stone fruits), e.g. peach
              Aggregate, e.g. raspberry
False fleshy fruits   e.g. banana

Most juices are derived from fleshy fruits and these may be conveniently subdivided for processing into three categories.

(a) fruits which are pulped and their juices removed by pressing, e.g. apple, berry fruits
(b) fruits requiring the use of specialised extraction equipment, e.g. citrus fruits, pineapples
(c) fruits requiring heat treatment before processing, e.g. tomatoes, stone fruits

In most large fruit processing operations the plant is usually dedicated to one type of fruit.

Citrus fruits are unusual because the outer skin or flavedo is rich in essential oils and other tissues such as the albedo or carpellary membranes contain substances that give rise to bitter flavours. The processing of citrus typically involves separation of these various components as an important principle and for certain products such as comminuted bases, the various components are recombined in different proportions (see Section 4.3.2).

The flesh of stone fruits is separated from the stones or pits, not only to facilitate ease of handling, but also because the stones are further processed to obtain both fixed oils and glycosides. Fixed oils, such as those from peach, have application in the cosmetics industry and glycosides may be used as a source of other natural flavouring ingredients such as benzaldehyde.

Another important aspect of the processing of fruit is the presence of pectins. These substances contribute to the viscosity of fruit juices and assist in the suspension of colloidal material and tissue fragments that make up its cloud.

When most fruits are pressed, pectolytic enzymes are released and these will, if not destroyed, clarify and or cause gelling of the juice. Rapid initial processing of freshly pressed juice is therefore an important factor in determining whether cloudy juice is obtained (in which case the enzymes must be destroyed by pasteurisation to at least 95°C) or clear juice, in which case enzymes are allowed to act and may be enhanced by the addition of synthetic enzymes in further quantity.

4.2.2 Soft Fruit Processing

Many different types of press have been developed for processing soft fruits, especially apples, on batch and continuous operation. The conventional pack press (Figure 4.2) sandwiches layers of pulped fruit enclosed in press cloths between slotted wooden racks. The layers of racks containing fruit are then subjected to mechanical or hydraulic pressure by an ascending ram working against the fixed frame.

4.2.2.1 Pre-processing.

Pectin in fresh apples is mostly insoluble and this enables fruit to be pressed immediately after cleaning and milling. One of the characteristics of stored fruit is that the pectin tends to become more soluble [1] leading to difficulties in pressing. Most apples are therefore processed in the United Kingdom in the period between September and December when the fruit is fresh.

For processing of soft berry fruits, there are similarities in the pressing but significant differences in the pre-treatment. Soft berry fruits cannot, for example, be subjected to washing procedures because of potential mechanical damage, and must therefore be subjected to higher standards of harvesting.

In some berry fruits, notably blackcurrants, pectinesterases are present that demethylate pectin to pectic acid. This in turn may react with calcium present to form calcium pectate which forms a firm gel, making the fruit more difficult to press and reducing juice yields significantly.

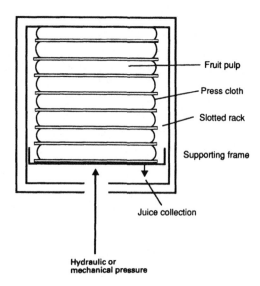

Figure 4.2 Diagram of a typical 'pack press' used for processing apples and soft fruits.

By warming the milled fruit to about 50°C and adding commercially available pectolytic enzymes that contain polygalacturonases, pectic acid can be broken down into galacturonic acid or its low molecular weight polymers before calcium pectate can be formed. By carrying out this process at around 45°C the reaction can normally be completed in 2–4 h to give a pulp from which juice may be freely expressed in good yield. The reactions involved are summarised in Figure 4.3.

In addition to the presence of pectins, many soft fruit juices contain

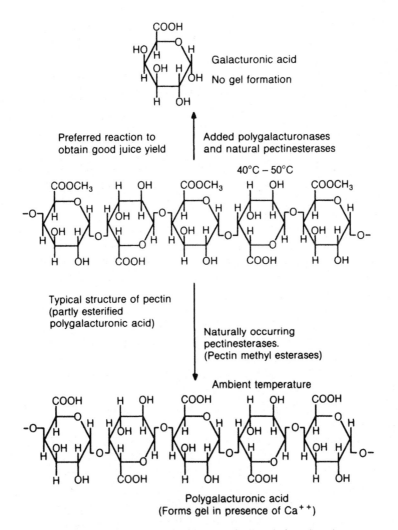

Figure 4.3 Reactions involved in enzymic degradation of pectin.

polyphenolic substances. These components may be involved in the formation of hazes and precipitates at almost any stages of the life of the products. They are also implicated in many non-enzymic browning reactions.

Polyphenols may, however, be removed by interaction with protein (e.g. gelatin). Hydrogen bond formation occurs between the phenolic hydroxyl group of the tannins and the carbonyl group of the protein peptide bond to give an insoluble complex. This reaction is affected by factors such as pH, concentration of polyphenols [2] and the type of protein used [3]. After formation the precipitate may be removed by filtration on filter aid or centrifugation.

4.2.2.2 *Juice processing after pressing.* Reference has already been made to the need for rapid pasteurisation of expelled juice. This process performs two functions; where cloudy juice is required it destroys pectolyic enzymes which would otherwise clarify the juice, and the naturally occurring yeasts and moulds that are a characteristic of all fruits and their untreated juices are largely killed off.

Various alternatives are available for further processing of expelled juice but typically, soft fruit juices are flash heated to boiling point and some 10–15% of the volume vaporised. Vapour and hot juice are separated in a cyclone with juice then passing to an evaporator where it is reduced to between 12 and 17% of its original volume (i.e. concentrated between six and eight times).

Separated vapour is normally passed to a fractionating column with reflux where the low boiling constituents giving the juice most of its aroma and flavour are separated from water which forms the bulk of the vapour. The volume of this fraction would be typically around 1% of the original juice volume and this fraction would be referred to as '100 fold volatiles'. This mixture of substances is a very important source of components for natural flavours (see Section 4.6.2).

As an alternative to evaporation and removal of volatile substances, the expelled juice may be filled into containers for the domestic market.

Some processors link de-aeration to the process of pasteurisation before concentration takes place. Dissolved oxygen has a degradative effect on juices and it has been assumed that this will have a rapid spoilage effect when the juice is heated. The de-aeration process when used, involves flowing the juice over a large surface area in a vacuum chamber where pressure is reduced to at least 100 mbar.

De-aeration is undoubtedly successful in removing air entrained in the product and this has the effect of reducing foaming in an evaporator. The link between dissolved oxygen and rapid spoilage during heating has not, however, been successfully proved [4].

Many processors do not find de-aeration to be of much practical value because of the lack of observed degradation and especially because of the loss of volatile substances that often results.

4.3 Specialised Fruit Processing

4.3.1 *Citrus*

Botanically, citrus varieties are forms of berry fruits in which the hairs inside the ovary walls form juice sacs. Epicarp is the familiar highly coloured, oil bearing outer layer. Both juice and oils are now valuable commodities in all citrus varieties and the recovery of both materials are important to the economics of processing.

Many different processes are used world-wide for citrus types with a two-stage operation being widely employed. In a typical process, the fruit passes over an abrasive surface or roller where the sacs in the epicarp are pierced and oil washed away by water spray. The resulting oil-in-water emulsion is screened to remove vegetable debris, oil separated by centrifugation and then dried and packed.

The rasped fruit then move onto an extractor where the juice is removed leaving albedo (pith) and peel (flavedo). Various juice extractors have been used with fruit being compressed in a roll mill or screw press, or the juice bearing material reamered out. Expressed juice is subjected to screening (sometimes referred to as finishing) before being further processed.

Other juice processing machines have been devised to minimise the handling of fruit and to incorporate both oil and juice extraction into a single operation. Of these the FMC 'in line juice extractor' is probably the best known. Its operation is based upon pairs of stainless steel cups with radiating fingers that intermesh when the upper and lower halves of the cup are brought together. Both cups are fitted with centrally mounted circular cutters; that in the bottom half is hollow and connected to a perforated tube beneath. A single fruit is fed automatically into the opened lower cup; the upper cup then descends forcing the fruit into the perforated tube from which the juice escapes. Residual albedo is collected separately.

Simultaneously, the peel is abraded and flaked by intermeshing fingers. These peel flakes contain the bulk of essential oils and when compressed in a screw press, yield an emulsion from which the essential oil may be collected by centrifugation.

Other pressing equipment has been developed for lemons which yield a more valuable oil and for which the demand for juice, although significant, is not as great as for orange.

Limes are normally processed in a slightly different manner. Washed fruit is compressed in a screw press to yield a pulpy juice that typically also contains the oil emulsion Larger pieces of pulp are screened out and in the classical process, juice and oil emulsion are fed to large tanks where a natural separation process occurs. Naturally occurring enzymes, which may be enhanced by the addition of commercially available synthetic pectolytic enzymes, clarify the juice whilst the oil bearing emulsion and pulp settle to the top of the tank. At the same time other debris settles below the clarified layer.

This process is normally completed in 10–30 days although in adjacent processing factories, there may be variations in time and juice clarity.

Clarified juice is then typically filtered and concentrated whilst the oil bearing emulsion is steam distilled. The process often requires a period of some hours of heating before actual distillation starts and it is in this way that the oil is brought to the specification required by the consumer. Steam distillation of lime oils usually brings about a number of changes to the components present in the undistilled oil.

Many variations of the classical process for limes have been developed with emphasis on the production of juice of low colour by ultra filtration and on the production of cold pressed oil by direct centrifugation of the oil bearing emulsion. Cold pressed lime oil is, however, of different character to distilled lime and is often used in different applications.

The fate of citrus juices both during and after processing will vary with their subsequent use. Whilst lime juice is normally available only as a clarified juice, lemon and orange juices are available both clear and cloudy. Because of the very high natural acidity (up to 8% as citric acid) of lime and lemon juices, clarification can take place using unpasteurised juices which will not normally ferment. Orange juice must be treated differently. Clarification is carried out by addition of pectolytic enzymes after pasteurisation to destroy microbiological activity.

As with soft fruit juices, the bulk of citrus juices are subjected to concentration to facilitate shipping and during this process, volatile components are usually collected separately from the juice concentrate. Both oil and water phase volatile fractions are collected from processes and these are widely used in flavourings.

4.3.2 Comminuted Citrus Bases

There is at the time of writing, a special category of soft drinks in U.K. food legislation [5] for which comminuted bases are an integral part. Some of the early processes which were developed in the first half of this century were patented and have been reviewed [6]. In an original process, whole oranges were shredded into sugar syrup which extracted both juice and oil emulsion. By screening the syrup, peel, seed and coarse rag were removed and the syrup pasteurised to yield the finished product.

More recent developments, which are almost always confidential, work on the principle of taking different parts of the fruit (except seeds) in varying proportions, blending them and milling the resulting mixture through a carborundum stone mill which 'comminutes' or disintegrates the solid components, releasing pectin and emulsifying the juice and oil components. The mixture must be rapidly pasteurised at 95°C to destroy enzyme activity, stabilise cloud and prevent microbiological spoilage.

Comminuted bases have a number of advantages for the soft drinks industry and, to a lesser extent, elsewhere in the food industry. The principal advantages are summarised as:

(a) a more intense flavour that is often preferred to the corresponding concentrated juice
(b) excellent cloud characteristics because of incorporated oil, pectin and other fruit materials
(c) good raw material stability

The main disadvantage of comminuted bases is in products exposed to oxygen, when the oil incorporated in the comminute base will often oxidise with corresponding characteristic flavour deterioration.

In more recent years, the boundary between cloudy juices and comminutes has become much less distinct. By incorporation of a small percentage of peel components or pulp extractives into juice it is possible to make significant cost reductions and whilst such materials may be satisfactory or even preferable for some manufacturing uses, they have found their way into pure fruit juice (particularly orange) at times of raw material shortage and corresponding high costs. This had led a number of European countries to set up sophisticated analytical testing to confirm juice quality.

4.3.3 Pineapple Juice

Pineapples are somewhat unusual because the process to manufacture juice is mainly a by-product of the pineapple canning industry. Whilst pineapple juice does have a characteristic flavour, the overall perception of this product relies heavily on the texture associated with the presence of tissue.

Pineapple is also typically low in pectin content and in consequence, enzymatic degradation of this pectin can result in rapid clarification. It is then essential in pineapple processing to ensure rapid deactivation of enzymes by heat treatment.

The major processing areas of the world are Hawaii and the Philippines although the relative ease with which pineapples may be grown and harvested means they are grown very widely in many developing nations.

In pineapple processing, the fruit must be peeled, cored and sliced. The cores, trimmings, small and outsize fruit are typically milled into a puree which is pasteurised in a tubular unit before being fed to screw presses which extract the juice. This juice will, depending upon the final quality required, then be screened or centrifuged before being passed to the evaporator for concentration or canning as single strength juice. Some products may be subjected to homogenisation to further stabilise the cloud components.

During evaporation to produce the concentrated juice, volatile components are normally collected for use with either the reconstituted juice or as flavour ingredients.

4.3.4   *Processes Requiring Heat*

Probably the best example of fruit products requiring heat treatment before processing into juice is the tomato although this process is normally applied to stone fruits as well. Tomato like pineapple juice is dependent to a large extent upon suspended solids to provide the characteristic flavour sensation and rapid enzyme deactivation by heat treatment is essential to avoid clarification.

In the process, cleaned fruit is usually coarsely chopped and heated in a tubular heat exchanger following which it is screened to remove peel, seeds and other debris. Secondary screening through paddle finishers with a mesh size of about 0.5 mm will normally give the typically pulpy juice.

Other processing may be used such as stone-milling or homogenisation to improve smoothness or cloud and a limited evaporation of water in scraped surface heat exchanges may be used to produce concentrated paste.

Stone fruits are often processed in a similar manner although juice production may, as with pineapple, be associated with waste and rejected fruit from a canning operation. The stones (or pits) from the fruit are subjected to drying and may be used in a further process as previously described.

Fruit processes involving heating may change the character of the flavour associated with the fruit and the flavourist must establish whether a 'fresh' or 'processed' character is required when using or matching these products.

4.4   Products and Packaging

A wide range of fruit products are available for the flavourist although their range of use is often affected by their degree of concentration and the packaging in which they are available.

As indicated previously almost all juice products must be pasteurised to control enzyme activity and the levels of microflora that are responsible for spoilage. Heat treatment is often incorporated as part of the packaging process although this is usually a second process to that carried out immediately after the juice is expelled.

Because of their low pH, most fruit juices and their concentrates will not support the growth of pathogenic micro-organisms and in consequence, the pasteurisation of products at between 80°C and 85°C for around 30 s is normally sufficient to remove spoilage organisms. In more recent packaging developments using aseptic containers ultra high temperature (UHT) conditions (e.g. 2–3 s at 120°C) may be used to sterilise the juice before it is packed in aseptic conditions to ensure asepticity. Low acidity juice especially tomato and some tropical fruits with higher pH values require more stringent conditions to give sterile products.

Most fruit products, whether concentrated or not, (except possibly the volatiles referred to earlier) are available commercially in a number of alternative packaging types.

4.4.1   *Frozen Juices*

Probably the largest commercial volume of trade in concentrated fruit juices is in frozen form. Orange juice concentrate accounts for the largest volume of frozen juice concentrates.

The process of freezing is not practically applicable to other than very concentrated (normally at least six times) juice because of the ice crystal/block size. With highly concentrated juices, the product generally remains as a firm slush at around $-20°C$ to $-25°C$. With a single strength juice the whole package would form into a single unmanageable block of frozen juice and cloud would probably be reduced. Typical packaging for frozen juice employs 200 litre steel drums with removable heads, the juice being held in two large polythene bag liners inside the drum.

Freezing of juices has many advantages as it maintains flavour and colour for long periods whilst keeping enzymic and microbial activity at a minimal level. The principal disadvantages of frozen juices are the practical problems of shipping and storage at around $-25°C$ (although the industry is now well able to accommodate these) and the need to allow adequate time (normally around 48 h) for thawing to ambient temperature to take place before the juice is used.

A further disadvantage is that some processors may, to enhance flavour, concentrate a product to a degree greater than that required and then to dilute or 'cut back' the concentrate to say a six times value with fresh unpasteurised juice. This will give the product an excellent flavour and appearance but may re-introduce active pectolytic enzymes and also microbiological contamination. These are held in check as long as the product remains frozen but upon thawing, gelation or fermentation may subsequently occur. If the frozen juice is to be diluted and packed for direct consumption it will have to be resubjected to a thermal process and this potential problem is dissipated. If however, the juice is to be used as a concentrated base for a flavouring, the problems indicated may arise.

4.4.2   *Aseptic Packaging*

Packaging developments over the past twenty years or so have led to aseptically packed juice in containers of sizes up to about 1000 litres becoming a technically and commercially alternative form of packing. The most familiar form of aseptic packaging for fruit juices is the 250 ml and 1 litre pack for domestic consumption and the principle can readily be applied to different pack sizes and to different degrees of concentration.

Pasteurisation of juice, either as a concentrate or at single strength is best carried out under UHT conditions (e.g. 2–3 s at 120°C) although satisfactory products can be achieved by using more conventional conditions such as 85°C for 30–45 s. This juice is carried to the aseptic filler, the key part of the operation, and filled into a flexible (and often multilayer) bag that is supported by some external rigid container (e.g. openhead steel drum, strong board case, etc.). The bag will have probably been sterilised by γ-irradiation and it is opened, filled and resealed in an aseptic atmosphere which may be steam, sterile air or a chemical spray such as hydrogen peroxide/peracetic acid.

Juice packed in this way will normally be microbiologically stable for an infinite time (unless the inner sterile container is opened or punctured for any reason). Commercially, this form of packaging is attractive because of ease and lower cost of storage and shipping. There is also an effective guarantee that the product will be aseptic at reconstitution and this ensures that the microbial load is minimal when further processing is started.

The principal disadvantage of aseptic packaging arises because juices packed in this way are usually shipped and stored at ambient temperatures, and browning frequently occurs. This may be a significant disadvantage for juice that is to be used for consumption as such; it may be less of a disadvantage when the juice concentrate is to be used as a flavouring component. Browning of colour is sometimes accompanied by unacceptable flavour changes.

4.4.3   Self-Preserved Juice

Many clarified soft fruit juices may be concentrated to around 68–70° Brix (see specifications) at which level of solids they are effectively self-preserving. Such juices may be, in some circumstances, vulnerable to spoilage by moulds; the growth of moulds usually occurs in closed containers subjected to heating and cooling. Water vapour from the concentrate condenses inside the container and drains back to the surface of the concentrate allowing the development of conditions to support mould growth. Self-preserved juices, which are particularly important for use in flavourings, are best stored in cool, 5–10°C, even temperatures in full containers. Occasional problems occur with osmotolerant yeasts which cause slow fermentation even at high solids levels. These problems are usually self-evident.

4.4.4   Preserved Juice

A frequent method of storing fruit juice concentrates is by the use of chemical preservatives although the Fruit Juice and Nectar Regulations [7] do not allow significant levels in juices for consumption as such. Preserved juices are probably most widely used in the preparation of bases for use in the flavouring and formulation of soft drinks. The presence of preservatives which may interact with other flavouring components may restrict other uses to which these products may be put.

Probably the most widely used preservative in concentrated juices is sulphur dioxide at levels of 1500–2000 mg/kg. In juice concentrates with high levels of natural sugars the amount of 'free' sulphur dioxide diminishes with storage. Sulphur dioxide reacts with many flavouring ingredients such as aldehydes, and any juice concentrate required for use as a flavouring base should not contain this ingredient.

4.4.5 Hot Pack Products

This process relies on raising the temperature of a juice to at least 70°C, usually by means of a plate heat exchanger, filling the hot product into its final container, closing the container and inverting it or otherwise heating the closure to pasteurising temperature. The pack is maintained at a required temperature to achieve pasteurisation and then cooled.

Relatively small amounts of juices are available in a sterile 'hot pack' form. This type of packaging is typically used for single strength (unconcentrated) juices where a high quality product is required. Most forms of packaging other than aseptic packs are inappropriate for unconcentrated juices and aseptic packs are rapidly displacing hot packs (usually cans).

4.5 Product Specification(s)

As indicated above, most juices and their components that are used in flavourings are concentrated in some measure because unconcentrated juices are unlikely to contribute much flavour except in a significant proportion (e.g. 5–10%) in the finished product.

At the time of writing, official methods of fruit juice analysis are under discussion within the EEC but none have been formally adopted. Methods published by the International Federation of Fruit Juice Producers [8] and the Association of Official Agricultural Chemists [9] are, however, widely used.

4.5.1 Soluble Solids Content

The solids content is a measure of the amount or degree of concentration and thus an important factor especially when fruit or juice content is to be claimed in a finished product. The most common analytical measure applied to juices of whatever concentration, is their soluble solids content. This is determined classically by filtering a known quantity of the juice to remove suspended solids and then evaporating the resulting solution at 105°C. The solids content is then determined by weight. This classical method is cumbersome and not regularly used; the most widely used routine method of estimating solids content is by the use of the refractometer.

The results obtained from the refractometric observations are based on Brix values for sucrose solutions. For juices with relatively low acidity, a reasonable relationship exists between actual soluble solids content and observed Brix value. The relationship is improved by the use of acidity correction tables.

For pure sucrose solutions, the relationship between the Brix value (%w/w) and the solids content in grams per litre (%w/v), which is generally of greater practical significance in experimental work, is a function of density. Thus

$$(\%\text{solids content w/w}) \times \text{density} = \%\text{solids content w/v}$$
$$(\text{g/ml } 20°\text{C})$$

These relative values are set out in Table 4.1. The relationship between observed Brix value and soluble solids content is much less satisfactory for highly acidic juices such as lime and lemon although the use of the refractometer for control purposes in any one factory remains valid. In

Table 4.1   Table of constants for sugar solutions at 20°C. B = °Brix (%w/w), GPL = grams per litre, SG = specific gravity

B	GPL	SG	B	GPL	SG
1	10.0	1.000	36	416.3	1.156
2	20.1	1.005	37	429.7	1.161
3	30.3	1.010	38	443.2	1.166
4	40.6	1.015	39	456.8	1.171
5	50.9	1.018	40	470.6	1.177
6	61.3	1.021	41	484.5	1.182
7	71.8	1.026	42	498.5	1.186
8	82.4	1.030	43	512.6	1.192
9	93.1	1.034	44	526.8	1.197
10	103.8	1.038	45	541.1	1.202
11	114.7	1.043	46	555.6	1.208
12	125.6	1.047	47	570.2	1.213
13	136.6	1.051	48	585.0	1.219
14	147.7	1.055	49	599.8	1.224
15	158.9	1.059	50	614.8	1.230
16	170.2	1.064	51	629.9	1.235
17	181.5	1.068	52	645.1	1.241
18	193.0	1.072	53	660.5	1.246
19	204.5	1.076	54	676.0	1.252
20	216.2	1.081	55	691.6	1.257
21	227.9	1.085	56	707.4	1.263
22	239.8	1.090	57	723.3	1.269
23	251.7	1.094	58	739.4	1.275
24	263.8	1.099	59	755.6	1.281
25	275.9	1.104	60	771.9	1.287
26	288.1	1.108	61	788.3	1.292
27	300.5	1.113	62	804.9	1.298
28	312.9	1.118	63	821.7	1.304
29	325.5	1.122	64	838.6	1.310
30	338.1	1.127	65	855.6	1.316
31	350.9	1.132	66	872.8	1.322
32	363.7	1.137	67	890.1	1.329
33	376.7	1.142	68	907.6	1.335
34	389.8	1.146	69	925.2	1.341
35	403.0	1.151	70	943.0	1.347

establishing the degree of concentration of a fruit juice, it is thus not possible to relate strictly the solids content, (degrees Brix) to this value. A 60° Brix orange juice is normally referred to as a 6:1 concentrate and this may be true for some juices. If, however, the typical observed Brix value of single strength juice (as obtained from the fruit) is 11.0°, then concentration is as follows:

	Brix (observed) %w/w	Solids content (g/litre) %w/v
Single strength juice	11.0	114.5
Concentrated juice	60.0	771.72
Degree of concentration		$\dfrac{771.72}{114.5} = 6.74:1$

These degrees of concentration are more significant when the fruit juice content to be added to a product is of special significance; e.g. to meet a legislative standard for juice content. For the more practical considerations of a juice concentrate that will give a good flavour base without undue risk of fermentation, then 60° Brix would normally be considered a minimum level (except for acidic juices such as lemon or lime).

Use of the hydrometer as a rapid method for estimating the soluble solids content for fruit juices should also be mentioned.

4.5.2 Titratable Acidity

The measurement of titratable acidity in a fruit juice is a very important indicator to the flavourist of the 'sharpness' of the juice. It is normally measured by direct titration with a standardised aqueous alkali solution (e.g. 0.1 N sodium hydroxide solution) using either a pH meter or phenolphthalein indicator to establish the end point.

For citrus and many other juices, acidity is calculated as citric acid (care must be taken to specify the result either in terms of monohydrate or anhydrous form) and results are conveniently expressed as grams per litre (gpl). Many juices contain acids other than citric (e.g. malic, oxalic etc.) and for products such as apple juice concentrates, results will normally be expressed in terms of malic acid (gpl).

For highly acidic juices, determination of acidity is the most important criterion in establishing the degree of concentration of the juice in a parallel way to that described above.

When consideration is being given to the use of a fruit juice concentrate as base for a flavour, acidity is important in establishing whether any (rapid) changes are likely to occur to the added flavour ingredients, natural or otherwise. Many reactions are catalysed or otherwise affected by acids and it is important to avoid changes in flavour character. For example, when high levels of some essential oils are compounded into acidic juice bases, there is often a rapid oxidative degradation of terpene constituents.

A further consideration to the use of acidic juice concentrates in compounded flavourings is the application for the flavouring. If, for example, a flavouring is to be added to dairy products, consideration must be given to whether the level of the acidity will affect proteins present. In some sugar confectionery applications, the presence of acids will cause some partial inversion of sucrose that may have a significant and, in some situations, undesirable effect on the finished product.

4.5.3 Brix Acid Ratio

The Brix acid ratio is a concept that has been widely used in the juice packing industry. It is an arithmetical proportion of soluble solids to (citric) acid content. It is used to indicate whether a juice is sweet, sharp, or 'median' in character and is also a useful indicator of the degree of ripeness of the fruit from which the juice originates.

Juice packers often specify the 'ratio' of the product required to ensure that there is minimal batch to batch variation. The ratio is calculated as follows:

$$\frac{\text{Observed Brix value}}{\text{Acid content}} = \text{Ratio}$$

It should be noted that the acid content used in calculating ratios, must be as %w/w and not %w/v. In this way, the result is not changed with increasing Brix.

4.5.4 Other Specifications

Depending upon the flavour applications to which a juice is to be put, various other specifications will be more or less important. The main factors to be considered with some comments on their significance, are as follows.

4.5.4.1 Routine quality factors

pH value: This may be important in determining the application to which the juice is to be put (see Section 4.5.2).

Colour: Juice concentrate colour is especially important for some flavour applications, especially red fruit flavours where the natural colour of the juice may be used as an (undeclared) colouring material. Deterioration of colour (browning) will also then become an important consideration.

Pulp (screened or suspended): The presence of pulp is very important in some fruit drinks as consumers often associate it with overall quality. In the manufacture of flavourings, pulp is normally unacceptable as it will usually cause difficulties in manufacture. It will therefore be important to specify the concentrate accordingly.

Oil content: In fruit juices other than citrus, no oil is normally present and in this respect, citrus juices are unique. In many citrus juices (and especially communites) there will often be residual or added oil and it will be most important to know how much oil is present in a compounded flavour.

Glycosides: In some fruit juices, especially citrus, glycosides may be present (e.g. naringin in grapefruit and hesperidin in orange). Many glycosides are bitter and may affect the character of the juice concentrate. Glycosides too may interact with other flavour components. Juices are not normally supplied with specified glycoside levels but methods are available for estimating them.

Ascorbic acid: Many juices have naturally occurring ascorbic acid and in some processing plants, ascorbic acid may be added as an anti-oxidant. In juice concentrate to be used as flavouring bases it will often be important to have a level of ascorbic acid to minimise anti-oxidant effects; 200–400 mg/kg would be a typical addition.

Preservatives: Most juice concentrates for flavouring manufacture will be required free from preservatives and this will normally not present any difficulties to the supplier apart from ensuring juice processing takes place before any microbiological spoilage occurs. Some juices (e.g. lemon and lime) are often supplied with high levels of sulphur dioxide to both prevent spoilage and minimise browning. As already indicated, sulphur dioxide will react with many added oil and flavouring components and it will normally be essential to ensure juices for use in compounding are free from this preservative. Other preservatives such as benzoic and sorbic acids are less likely to interact with flavour components but are best avoided in juice for flavouring use as they both have a distinctive flavour characteristic.

4.5.4.2 *Stability criteria*

Viscosity: The viscosity of a juice may not be an important factor in most flavour applications although if other components are to be emulsified into the juice, viscosity may increase still further to a level which is unacceptable. Increasing viscosity may, however, be an early indicator of gelling (see below).

Separation: This factor only applies to cloudy juices but may be important as no separation will normally be acceptable in the finished flavouring. Separation of the juice or concentrate may also be indicative of deteriorating or ageing juice.

Turbidity: Turbidity measurements, usually carried out after centrifugation of a cloudy juice, will indicate the amount of cloud present. Cloud levels in a juice will normally be more important for applications and uses other than flavouring manufacture.

Gelling and enzyme activity: The level of sugar in many juice concentrates is often ideal for gel formation to occur. The occurrence of gels is normally associated with pectolytic enzyme activity at some stage during the juice manufacture. Methoxyl groups are removed from pectin by the activity of the enzyme pectinesterase and in this state, gel formation may occur on storage, particularly in the presence of calcium ions (see Section 4.2.1).

Gelling is undesirable for most applications and tests are available to establish the tendency of the concentrate to form a gel. Formation of a gel indicates usually either delay in processing pressed juice or inadequate thermal processing. The effect may not become apparent until the juice has been stored for some time and may also indicate a low level of residual enzyme activity.

4.5.4.3 *Microbiological evaluation.* Although the addition of many flavour ingredients to juice concentrates will have the effect of preventing further spoilage occurring and will thus render the juice based flavouring stable, it is essential to ensure that the juice is of good microbiological condition before manufacture is undertaken. Tests will normally be established for viable counts at 20°C and 37°C and espcially for yeast and mould counts. If yeasts are present, the presence of osmotolerant yeasts can cause problems in concentrates. The presence of some acid tolerant organisms may also cause problems (e.g. development of diacetyl and acetylmethyl carbinol).

Of particular significance in fruit juice concentrates, is the level of contamination by mould spores. Not only can they cause difficulties in concentrates before the product is used (e.g. flavour deterioration), but even in a compounded flavouring containing juice concentrate, the mould spores may, in some circumstances, survive to contaminate the final foodstuff in which the flavouring is used.

4.5.5 *Juice Adulteration*

The adulteration of fruit juices may be of greater concern to the packer intent on selling the juice as such than to the flavourist. In some circumstances it may be important to the flavouring compounder to know whether the juice being used has been adulterated as the possibility of unwanted reactions between possible adulterants and flavouring ingredients may have to be considered. Alternatively, the use of a lower cost juice that could not be used for consumption as such, may be acceptable for a flavouring application.

The following analytical checks are some of those widely used to indicate possible juice adulteration in citrus juices and may have wider applications in other juices. In detection of adulteration it is vital to establish a wide database as the key parameters of a juice may vary naturally from country to country and from season to season within a country. In order to establish the probability of adulteration, mathematical methods are used to compare the parameters.

Unless adulteration is of a gross nature, it may be extremely difficult to detect objectively without access to a method such as isotope analysis. Some of the parameters important in establishing adulteration are as follows:

spectrophotometric examination
phosphorus
citric acid/isocitric acid ratio
polyphenolic substances
sugar analysis
reducing sugars
amino acids (total, by formol titration)
spectrum of individual amino acids
protein content
ash

Individual values of the above parameters may not have particular significance. Comparative values of all the parameters, particularly against historical data are important and the experienced analyst can usually readily identify an adulterated juice.

Table 4.2 Typical comparative data on the composition of fresh orange juice

Parameter	Units	Europe [11]	United States [10]
Relative density	20°/20°C	Min 1.045	Mean 1.046
Brix corrected	°Brix	Min 11.18	Min 10.5
Volatile acid (as acetic)	g/litre	Max 0.4	—
Ethanol	g/litre	Max 3.0	—
Lactic acid	g/litre	Max 0.5	—
Sulphurous acid	mg/litre	Max 10	—
Titratable acidity	meq/litre	106–160	—
Citric acid	g/litre	7.6–11.5	6.3–19.7
D-Isocitric acid	mg/litre	65–130	—
Citric/isocitric	Ratio	Max 130	—
L-Malic acid	g/litre	1.1–2.9	1.05–1.77
L-Ascorbic acid	mg/litre	Min 200	270–870
Ash	g/litre	2.9–4.8	2.8–7.3
Potassium	mg/litre	1400–2300	1200–2770
Sodium	mg/litre	Max 30	2.0–25
Magnesium	mg/litre	70–150	102–178
Calcium	mg/litre	60–120	65–305
Nitrate	mg/litre	Max 10	—
Total phosphorus	mg/litre	350–600 (as PO_4)	84–315
Sulphate	mg/litre	Max 150	—
Formol index per 100 ml	ml.1 N NaOH	15–26	12–30
Hesperidin	mg/litre	Max 1000	835–1230
Water soluble pectin	mg/litre	Max 500	—
Glucose	g/litre	20–28	—
Fructose	g/litre	22–30	—
Glucose/fructose	Ratio	Max 1	—
Sucrose	g/litre	Max 45	30–65

To the flavouring compounder, perhaps the most likely compounds of interest are the reducing sugars and amino acids which could enter into Maillard reactions. Aroma and flavour of the juice remain the ultimate consideration.

Typical composition values of fresh orange juice are shown in Table 4.2.

4.5.6 *Specifications for Essence/Volatiles/Citrus Oils*

Because of the particular significance of juice volatiles and citrus oils in the manufacture of flavourings, there are a number of specifications and analytical parameters that should be considered in the use of these raw materials. As in all flavour ingredients, the ultimate specification is the subjective evaluation of aroma and flavour by experienced tasters. Some may also consider the gas chromatography profile to be of particular significance.

4.5.6.1 *Aldehyde content.* This characteristic is of particular importance in lemon and orange oils where it is normally calculated as citral and decanal, respectively, as a measure of the aromatic character of the oils. As with many such measurements, the sample to sample variation is probably more significant than an isolated specific value.

4.5.6.2 *Refractive index.* This measurement is widely used in the specification of flavour ingredients and essential oils and is often a good indicator of the change in composition. For accurate measurement, the refractometer should be used at a constant 20°C and monochromatic light from a sodium source should be used as the source of illumination.

4.5.6.3 *Optical rotation.* This physical property of substances containing asymmetrically substituted carbon atoms is widely used although more in the evaluation of essential oils than flavouring ingredients. A beam of polarised light usually from a sodium lamp source is passed through a tube containing the substance under examination. Rotation of the beam to the right (dextrorotation) indicated by a + value or to the left (laevorotation) and indicated by a − sign is then used as a characteristic of the sample. The quantitative value is in degrees of rotation but reported for a tube containing 100 mm of the liquid under examination. If shorter or longer tubes are used the value is corrected back to 100 mm.

4.5.6.4 *Residue on evaporation.* This characteristic is not widely used in the evaluation of fruit essential oils and volatile fractions, but may be useful in detecting the additions of adulterants.

A weighed quantity of oil is heated on a steam bath for 5 h followed by 2 h in an oven at 105°C. During the heating it is possible to detect by smelling, unusual volatile substances and any residue will often indicate whether adulterants are present.

4.5.6.5 *Organoleptic quality.* This is possibly the most important if subjective evaluation of a fruit oil or essence as it is this characteristic that determines the use to which the material may be put. In the case of fruit juice volatiles, a 1.0-ml sample is added to 100 ml of a solution containing 125 g/l of sucrose and 2 g/l of citric acid. The final solution is tasted against a standard, and evaluated both qualitatively and quantitatively.

In the case of citrus oils, 1.0 ml of a solution of 2% of the oil in either ethanol or isopropanol is added to 100 ml of the above sucrose/citric acid solution and again evaluated by taste.

4.5.6.6 *Peroxide value.* This value is of significance for citrus oils generally as an indicator of the amount of deterioration undergone by the oil. It is often an indication of its age.

4.5.6.7 *Ester content of volatile fractions.* The ester content of the volatile fractions of fruit juices is determined by converting the esters to hydroxamic acids which are then reacted with the appropriate reagent and determined colorimetrically.

The above characteristics may not individually be of great significance but as with analytical parameters for juices themselves, may be used to build up a clear picture of the juice volatile or citrus oil. Against a databank of historical information they will, in experienced hands, give surprisingly clear information about origin, variety and adulteration, etc.

The above measurements are not an exhaustive list and the flavourist may well wish to consider other examinations such as the ultraviolet absorbance of citrus oils and the profile by gas liquid chromatography in order to broaden the picture still further.

4.6 Volatile Components of Fruit Juices

4.6.1 *Production*

So far this chapter has dealt with the production of different kinds of juices and some of their characteristics. A vital aspect of the selection and processing of fruit juices for use as flavourings is the isolation and use of the volatile components. In the case of citrus where oil components are also present the volatile components are usually subdivided into oil phase volatiles and water phase volatiles. In most other fruits the volatile substances are obtained only from the concentration of the juice and are thus known as water-phase volatiles.

These volatile fractions of juices have long been recognised as containing a major part of the distinctive aroma and flavour of a juice but it was not until 1944 that a successful process was developed in the United States for their recovery from the concentration process. The equipment then known as an 'essence recovery unit' is now commonplace in many of the juice processing plants around the world. A typical aroma recovery unit is shown in diagrammatic form in Figure 4.4.

As indicated above, a typical aroma or essence obtained from a fruit juice contains a large number of individual substances and as techniques for their deletion and recognition become more sophisticated, the number increases. The number of substances present may vary from 50 or less to several hundred with typical levels ranging from 1 ppb to 50 ppm. Control of the variation of

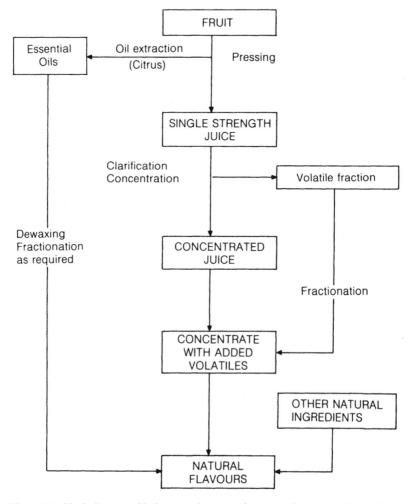

Figure 4.4 Block diagram of fruit processing operation to produce natural flavourings.

these components is a major problem in the use of fruit juice volatiles (see Section 4.7).

Many individual components have a low solubility in aqueous systems and when concentration occurs to the typical '100' fold volatiles, physical separation may sometimes occur. However, solubility of components in the concentrated volatile fraction is usually enhanced by the presence of ethanol, which, after water, is frequently the major component of juice volatiles. Ethanol levels can range from as low as 1% v/v to 15% or even 30% v/v in extreme cases. The amount of ethanol is often a useful indicator to the quality of the juice prior to concentration as it normally arises from fermentation by yeasts of the natural sugars in the whole juice.

Most volatile fractions would be expected to contain at least 1% of ethanol and values up to 5 or 6% would be considered both normal and acceptable. Above 6%, many consider that the amount of fermentation that has occurred will have materially affected the characteristics of the aroma although the net result may still be acceptable to the flavourist. From a commercial standpoint, ethanol content in volatile fractions can have a dramatic effect on costs in countries (like the United Kingdom) where levels of excise duty are high. There can often be a significant commercial benefit in reblending the volatile fraction with concentrated juice before importation as this can usually be arranged to reduce ethanol levels to below the threshold of excise duty. After ethanol, all the aroma substances together rarely exceed 0.5% of the concentrate.

Stability of the various volatile fractions from different fruits may vary considerably although this can be significantly influenced by the choice of plant design. In an appropriately designed unit, thermal stability of volatiles is usually excellent although even in the most sophisticated plant, substances such as ethyl acetate will usually undergo at least partial hydrolysis and this can result in changes in the aroma if the hydrolysed esters have a significant effect on the flavour.

An important aspect of the collection of juice volatiles is whether or not their components have any tendency to form azeotropic mixtures with water. Walker [12] has suggested a classification of fruit juices in this respect.

Class I juices contain flavour constituents that exhibit strong azeotropic properties (e.g. Concord grape, pineapple). Class II contains important flavour constituents exhibiting weak azeotropic properties (e.g. strawberry). Class III contains no important flavour ingredients that exhibit azeotropic characteristics (e.g. apple, pear).

For the design of equipment to collect volatile fractions which exhibit any azeotropic properties it is necessary to have information about the azeotropic equilibria between the volatile substances and water, as well as with juice. It also follows that a system designed to collect volatiles from a specific juice may not necessarily perform well with other juices. This may be important in determining the quality of volatiles for flavouring uses.

When juices are processed through a volatile recovery system it is, under some circumstances, possible to cause chemical changes to the non-volatile residues (the concentrated juice) that cause undesirable changes in colour (browning) and the production of unwanted cooked taste characteristics.

In early processes, recovery of volatile components involved heating the juice with between 5 and 50% of the volume being flash-evaporated at atmospheric pressure. Vapour removed in this way was then fed to a distillation column where rectification took place to yield an 'essence' of about 1/150th of the volume of the feed juice. Current techniques involve flash distillation of a juice under vacuum to remove volatile substances before concentration of the non-volatile ingredients to the required level (also under vacuum).

The volatile substances (or essences) are not only used for the manufacture of natural flavourings but also for re-addition to concentrated juice such that on re-constitution, more of the aroma and taste of the original juice is recreated. This effect is also achieved by over-concentrating a juice and adding single strength juice to the required level of concentration. Such a product when re-constituted also shows flavour characteristics that are greatly enhanced compared with vacuum concentrated products.

4.6.2   Composition of Fruit Juice Volatile Fractions

The broad composition of the volatile fractions of fruit juices was referred to in Section 4.6.1. The qualitative composition of the actual volatile substances is

Table 4.3   Typical components of juice volatiles

Alcohols	Acids	Carbonyl compounds
Methanol	Formic acid	Formaldehyde
Ethanol	Acetic acid	Acetaldehyde
n-Propanol	Propionic acid	Propanal
Isopropanol	Butyric acid	n-Butanal
n-Butanol	Valeric acid	Iso-butanal
2-Methylpropan-1-ol	Caproic acid	Iso-valeraldehyde
n-Amyl alcohol	Caprylic acid	Hexanol
2-methyl butan-1-ol		Hex-1-en-2-al
n-Hexanol		Furfural
		Acetone
		Methyl ethyl ketone
		Methyl propyl ketone
		Methyl phenyl ketone
Esters		
Amyl formate	Methyl acetate	Ethyl acetate
n-Butyl acetate	n-Amyl acetate	Iso-amyl-acetate
n-Hexyl acetate	Ethyl propionate	Ethyl butyrate
n-Butyl butyrate	n-Amyl butyrate	Methyl isovalerate
Ethyl-n-valerate	Ethyl caproate	n-Butyl caproate
Amyl caproate	Amyl Caprylate	

now covered in greater detail here. As previously indicated, a large number of substances are usually present in the volatile fraction obtained during concentration of fruit juices. The substances found are principally esters, alcohols, carbonyl compounds and free (volatile) acids. Most of these substances are of relatively simple structure and a list, not exhaustive, of the commoner components found in many juice volatiles appears in Table 4.3.

The pattern of substances found in the volatile fraction of a juice is characteristic of a number of factors.

4.6.2.1 *Components that are characteristic of the genus.* Clear differences occur in the pattern of volatile substances that differentiate for example, apple juice volatiles from those of pear or blackcurrant juices.

4.6.2.2 *Components characteristic of the species.* Marked differences occur in the pattern of volatile substances that may be obtained from different varieties of fruits.

4.6.2.3 *Components characteristic of processing methods and/or technology.* Quite large differences may occur if the same variety of fruits are processed in different ways in the same factory or in different factories using various technological alternatives.

It has already been indicated that if a juice is allowed to stand for some time between pressing and pasteurisation, naturally occurring yeasts that are present on the fruit will begin to ferment the juice. This not only gives rise to the development of ethanol but also to a range of substances that are characteristic more of the fermentation products than of the juice itself.

Where relatively large amounts of ethanol develop, further infection, by, for example acetobacter, can in the right conditions, lead to the production of significant amounts of volatile acids, especially acetic acid. The presence of these two substances in some quantity can lead to further amounts of esters being produced or the transesterification of esters already present.

Further changes may occur within a specific concentration plant.

With all the possible areas for variation of the volatile components in a fruit juice, the flavourist may feel that it is almost impossible to obtain a consistent quality of volatile fraction for use in a formulation. In, practice however, individual processing plants are able to exercise a high degree of consistency.

Control of quality of volatile fractions of juices is almost always achieved by a combination of methods. Gas chromatography (GC) is often employed to give a 'fingerprint' of the active flavour ingredients present although the final selection and use of a consignment of fruit juice volatiles is by subjective evaluation of aroma and flavour. Aroma, as with other aromatic substances, is typically evaluated by use of the 'smelling strip'; an absorbent strip of firm paper which is dipped into the mixture of volatile materials and then smelt at regular intervals. Flavour is typically evaluated by experienced taster(s) using a solution of 12.5% sucrose, 0.25% citric acid anhydrous and up to 0.5% of the mixture of volatile substances.

4.7 The Use of Fruit Juices in Flavourings

4.7.1 Fruit Juice Compounds

This chapter is not devoted to the application of ingredients but the use of juice concentrates in the manufacturing of fruit juice compounds for use in the preparation of juice containing beverages; an extremely important sector of their use. In the manufacture of a beverage, typical constituents in descending order of quantity would be water, sugar, fruit juice, carbon dioxide, citric acid, flavouring, ascorbic acid, colouring and in some cases, preservative.

The function of a compound is to pre-mix all the required components, other than sugar, water and carbon dioxide, in a concentrated form. This enables the bottler to purchase a compound to which he adds sugar, water and carbon dioxide in order to obtain the finished beverage.

A 20-fold compound for a 10% juice drink would, for example, mean that by the addition of appropriate amounts of the missing ingredients, 1 litre of compound would contain sufficient of the key components to make 20 litres of finished beverage. A typical composition of the compound would be:

Orange juice concentrate 60° Brix	300 litres
Citric acid anhydrous	50 kilos
Natural cloudifier	20 litres
Natural orange flavouring	20 litres
Colouring	q.s.
Ascorbic acid	q.s.
Preservative	q.s.
Water to	1000 litres

Thus the compound would contain 300 litres of juice at 60° Brix (0.772 g/litre of solids), i.e. 231 g/litre of orange juice solids.

Assuming single strength juice to contain 115 g/litre of solids, 1 litre of compound would contain $(231 \times 100)/115 = 200\%$ juice solids. This diluted by a factor of 20 would give 10% juice solids in the finished drink.

The market span for beverage compounds in the United Kingdom is probably greater than in almost any other country in the world. This arises from the complexity of U.K. Soft Drink Regulations which allow for both concentrated soft drink and ready to drink versions.

A particular category in both strengths is for products containing comminuted citrus bases, the so-called whole fruit drinks. A variety of appropriate compounds are available in different strengths with all the ingredients necessary except water, sweetener and where appropriate, carbon dioxide. Comminuted citrus bases (see Section 4.3.2) contribute fruit to this product but also contain dispersed essential oils to enhance the flavour as well as peel and albedo components to enhance cloud and stabilise the overall beverage.

4.7.2 Flavourings

Fruit juice components are widely used in flavourings as such. They are the mainstay of natural flavourings of many types. The composition of individual formulations are confidential but typical recipes would be based on a high proportion of a fruit juice concentrate to which would be added appropriate juice volatile fractions as well as individual flavour components. For natural flavourings, these individual ingredients would be separated normally by physical processes such as fractionation.

A surprising degree of control may be exercised by the use of these techniques although, as indicated, consistency of supply from selected processing plants is an essential component of consistent flavour. A list of the typical components of fruit juice volatiles appear in Table 4.3.

4.7.2.1 The use of fruit juices as flavouring components.
Single strength fruit juices are not widely used as ingredients of flavourings although concentrates and fractions obtained during processing are very important in terms of the quantity used.

Fruit juices, and to some extent purees and pulps are, however, often used in various finished food and beverage products as part of an overall flavour system. Whole fruit is an essential part of the manufacture of jams and conserves; it is impossible to distinguish the effect, in products of that nature, of the fruit itself and the flavour it brings to the product. Fruit juices are nevertheless added to products for a number of reasons. From a marketing standpoint it is often valuable to be able to claim that a product contains a certain percentage of fruit juice.

A related reason is the enhanced nutritional status that fruit juice content brings to a product. Many authorities are critical of some products, e.g. flavoured soft drinks which are described as having empty calories. The addition of fruit or fruit juice largely dispels such criticisms. The product with say, a 10% juice content is seen as having a significant nutritional benefit when compared with simply flavoured products.

The addition of fruit juices to food and drink products whether beverages, ice cream and other dairy products, or other foods, is almost always by use of concentrates appropriately reconstituted. Single strength juices are rarely used in products because of the very high cost of transporting them to the point of use and the volumetric problems often posed by most aspects of their use.

4.8 Summary

Fruit juices are normally considered in their own right or at least as direct ingredients of food or beverages and much of this chapter is relevant to them in any of these applications as well as ingredients of flavourings. There is much published on fruit and fruit processing although even this is sparse when their economic importance is taken into account. Much fruit juice technology remains in the hands of engineering companies and those associated with growing and processing fruit and is often considered commercially sensitive. The determined researcher can normally have access to most of the relevant data in a given area of juice technology but it remains surprisingly difficult to find a collected concise source of relevant basic data on the subject.

This chapter has therefore, examined some of the more important methods of manufacture of fruit juices, their composition and some methods of analysis. It has outlined the main forms of packaging used and parameters for specifications. The objective has been to provide the flavourist, applications technologist and others involved with flavourings with the basic information to enable a more systematic approach to be made to the procurement, evaluation and use of fruit juices. The information is slanted towards juices for use in flavourings but is equally relevant in most cases to other juice uses.

5 Synthetic ingredients of food flavourings

H. KUENTZEL and D. BAHRI

5.1 General Aspects

5.1.1 Introduction, Definitions and Documentation

The need for synthetically prepared flavour compounds arises from the fact that during the storage of foodstuffs a certain loss of flavour is inevitable. These losses can be compensated for by adding synthetically produced flavour compounds. Besides this, synthetic flavour compounds have the great advantage of being available in the required quantity and quality irrespective of crop variation and season.

A constant quality permits standardisation of the flavourings. Synthetically produced flavour compounds make it possible to vary the proportions of single components and thereby create new flavour notes.

In a description of this vast field, some definitions are essential. All substances added to a foodstuff are called ingredients, i.e. flavours, colours, emulsifiers, salt, etc. We distinguish between natural and synthetic ingredients according to their origin. In this chapter we concentrate only on flavour-enhancing or -modifying ingredients, and more specifically on the synthetic substances. Equally useful to circumscribe this subject are the definitions of IOFI (International Organization of the Flavour Industry) for natural, nature identical (synthetic) and process flavours (see Chapter 1). The anticipated new guidelines of the EEC will also be of great importance for future flavour definitions (see Section 5.3).

To illustrate the relationship between sensory perception such as flavour, taste and mouthfeel on the one hand and parameters related to the chemical structure such as volatility and molecular weight on the other, some examples of ingredients are listed in Table 5.1. In this chapter we concentrate mainly on volatile flavour components, thus omitting the polar and non-volatile ingredients which are responsible for taste and mouthfeel.

There are today several thousand flavour compounds known and described in a multitude of books, papers, patents, etc. and an efficient way of documentation is therefore mandatory. To fulfil the different needs of users, a number of specialised documentation systems were created. A most valuable tool is *Volatile Compounds in Food*, a compilation of about 5500 compounds found in 255 food products. This list edited by the TNO-division for nutrition

Table 5.1 Classification of food ingredients

Volatility	Organoleptic function	Molecular weight	Examples
High	Flavour, taste	< 150	Esters, ketones, simple heterocycles
Low	Taste, flavour	< 250	Carbonic acids, amids, extended C-skeletons
None	Taste	$50 \simeq 10\,000$	Sugar, amino acids, SMG, salt, nucleic acids, bitter agents, natural sweeteners
	Mouthfeel	> 5000	Starch, peptides
	Unknown	$> 10\,000$	Biopolymers, melanoids

and Food Research (Zeist, The Netherlands) is annually updated by a supplementary volume [1].

Another regularly updated work is the so-called GRAS-List, a world-wide reference list of materials used in compounding flavours and fragrances with sources of supply and data of the legislative status of each compound [2].

A number of classical books listing flavour and fragrance compounds, their use and description, should also be mentioned here. This list, by no means exhaustive, contains such well-known names as Arctander [3], Fenaroli [4], Gildemeister [5] and Guenther [6]. In recent years several authors have covered the whole field of ingredients and composition in a comprehensive manner [7–11].

Another source of information on synthetic flavour components are the catalogues of the flavour compound producers (a list of which can be found in [2]). In addition to the information in these sales catalogues a wealth of know-how and research results can be retrieved from the appropriate patents of these companies. The patent literature can be found in Chemical Abstracts, a documentation system which will be further explained below.

A review work covering a large range of relevant results is edited by CRC Press. In the periodical entitled *Critical Reviews in Food Science and Nutrition*, each year about a dozen articles summarise a topic of the indicated field [12]. For more than 40 years, Bedoukian has written a concise annual review article about perfumery and flavour materials [13].

A very extensive information system (including hard copies, micro films or electronically stored data) is offered by Chemical Abstract Services. It is possible to retrieve in an efficient way, specific data concerning synthetic, analytical or physical results. A search can be initiated by chemical formulae, key words, name of author(s), date of publication, etc. The possibility of combining any of these questions further enhances the value of this data system [14].

Additionally, there exists a broad variety of journals covering scientific and applied aspects as well as marketing and business questions of foodstuffs.

5.1.2   Flavour Generation

To gain a better insight into food flavourings it is important to understand the generation of flavours in foodstuffs. Four methods of formation can be distinguished:

enzymatic
non-enzymatic
fermentative
autoxidative
} formation of flavour compounds

In Section 5.2.3, some examples of each type are covered in more detail but a more general overview has to suffice here. With the enzymatic formation it is important in the first instance to know from which metabolic cycle the flavour compounds in question originate. In this context the most important metabolic cycles are those of fatty acids, amino acids, carbohydrates and terpenoids (the latter being secondary metabolites). Some general examples will further explain the situation.

The biosynthesis of fatty acids starts with acetic acid in an activated form. Chain elongation is effected by several enzymatic steps within the fatty acid-synthetase complex, adding a C_2-unit in each synthetic cycle, thereby producing even-numbered carbon chains. Hydrolysis and decarboxylation of the intermediate β-keto-ester generates the corresponding methyl ketone with one carbon less (see Figure 5.1a). These ketones may be the starting material for the corresponding secondary alcohols, which in turn may be esterified.

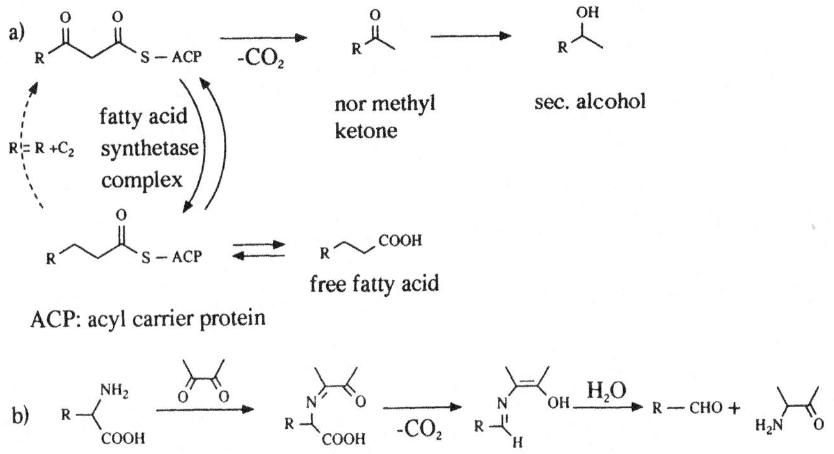

Figure 5.1   Generation of flavour compounds by different metabolic pathways: (a) from fatty acids [15, 16]; (b) from amino acids (Strecker degradation) [17].

Fatty acids are generally accepted as the origin of these flavour-active compounds [15, 16].

A simple and generally occurring transformation of amino acids to aldehydes with one carbon less is the so-called Strecker degradation. This reaction needs a carbonyl compound as counterpart for the transamination. α-Diketones are especially suitable for this sequence and give rise to α-aminoketones, which in turn are the origin of further reaction products (see Figure 5.1b). With radioactive labelling the relationship between the aldehyde formed and its parent amino acid can be proven as exemplified in Section 5.2.3.5. The so-called Strecker aldehydes are very important flavour compounds, either as volatile constituents themselves or as reactive intermediates for further transformation.

A very large and well investigated field within enzymatic flavour formation is that of the biosynthesis of terpenoids. The basic building blocks for the hundreds of known terpenoids are the 'isoprene units' shown in Figure 5.2. This C_5-unit is generated from mevalonic acid, which in turn is formed from three acetic acid units with every single step catalysed by a specific enzyme. The activated form of the C_2-starting material is a thioester of acetic acid with Coenzyme A (not shown in the figure). This is similar to the starting point of the biosynthesis of fatty acids. A pyrophosphate group (abbreviated as OPP) is the activated 'leaving group' for the OH group used by nature. Decarboxylation of mevalonic acid drives the reaction to the unsaturated C_5-unit, which readily undergoes further reactions. Fusion of two or three such units in a head-to-tail manner gives rise to a carbon chain with a characteristic array of the extra methyl groups. Their pattern can be found in a plethora of

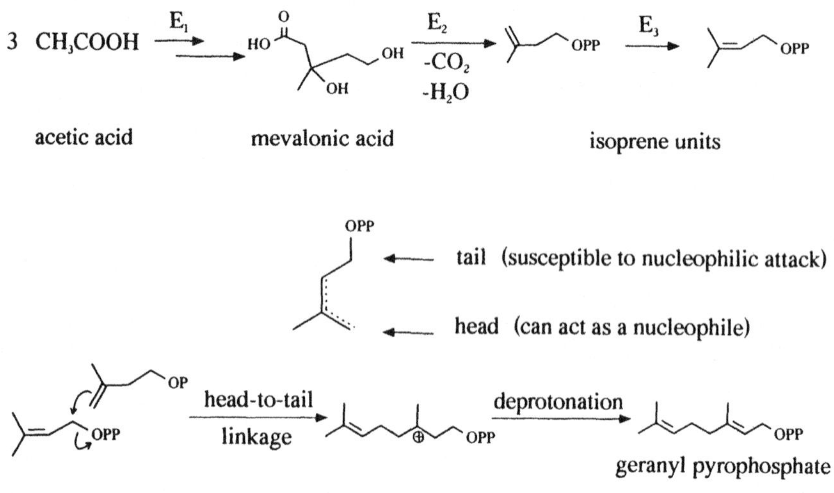

Figure 5.2 The first steps in the biogenesis of terpenes. E, enzyme or enzyme system; OPP, pyrophosphate.

monoterpenes (C_{10}) and sesquiterpenes (C_{15}), acyclic or cyclised, oxygenated or reduced, and is known to generations of chemists as the 'isoprene rule' [18, 19].

Besides these 'regular terpenes' a wealth of irregular compounds are formed, either by a Wagner-Meerwein type rearrangement or by fusion of the C_5-units in another way to the normal head-to-tail linkage. In Sections 5.2.3.3, 5.2.3.4 and 5.2.3.5 and in Figures 5.14, 5.17 and 5.20 some common examples of the biosynthesis of terpenoids are outlined. It is far beyond the scope of this article to give an overview of terpene biochemistry and reference to some general and review articles is sufficient [20–22a].

The generation of flavour compounds by a Maillard reaction (or a caramelisation) is described as non-enzymatic formation. The Maillard reaction is characterised as the thermal reaction of a reducing sugar with an amino acid. Thereby an overwhelming number of substances are formed. Their range covers the simplest degradation products such as H_2O, NH_3, H_2S, together with typical flavour compounds such as furans and pyrazines as well as the highly complex brown pigments known as melanoidins.

Characteristic of this type of reaction is the rich variety of substances formed (a variety related to chemical classes as well as to homologous series) and the very low concentration of the single compounds (usually in the range of parts per million).

In the formation of substances by thermal processes during a non-enzymatic reaction, the only reactions to occur are those generally considered in a chemical sense as simple (e.g. eliminations, additions, isomerisations). The above introductions may be helpful to explain the formation of many of these flavour compounds, but are not adequate as a guide for their practical preparation. The main reason is the low specificity of such thermal reactions. Chemical syntheses are normally expected to proceed in a way that yields are several orders of magnitude higher than in the undirected manner of these thermal processes.

The exploration of the most abundant browning reaction goes back to 1912, when Maillard investigated this process [23]. Since then many researchers have worked on the reaction and nowadays the first steps are fairly well known, but consecutive steps await further elucidation [24–26]. In Figure 5.3 a general scheme and a specific example of Hodge are given [27]. The amine function of proline reacts with the carbonyl function of glucose and rearranges to the Amadori compound 1-desoxy-1-prolino fructose. In this model reaction system, a number of cyclic enolones could be identified, all containing the C_6-skeleton of the starting glucose.

It is easy to imagine, that with a more complex mixture of starting materials (several sugars and amino acids) or under a higher reaction temperature, a very large range of products can be formed, partly by further degradation to smaller fragments, partly by incorporation of heteroatoms like nitrogen and sulphur.

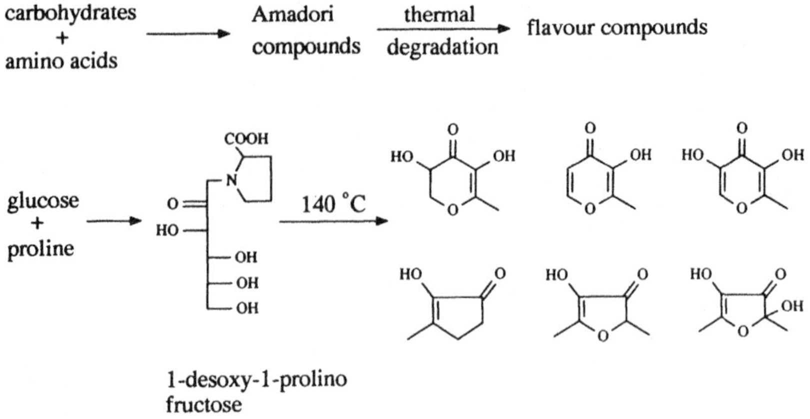

Figure 5.3   The Maillard reaction: general scheme and an example [27].

Less well-known but nevertheless an important part of Maillard products are the 'meta-stable' flavour compounds. Their restricted life-time reflects the familiar time-dependent increase and decrease of flavour sensations during roasting, baking and so on. The caramelisation reaction is the transformation of sugars or carbohydrates by strong heat treatment. Volatile components are important for the flavour whilst non-volatile substances form the brown colour. The structure and formation of the products responsible for the desired effects of flavour and colour are well documented [28, 29].

Fermentative formation covers classical processes such as the ripening of cheese, fermentation of wine and modern biotechnology. The old processes are based on ancient experience of mankind and probably also on trial and error. The modern processes of biotechnology, however, allow the production of high value products in a highly specific way. Biotechnology is a very fast growing science with an increasing influence on future food preparation and flavour generation [30, 31].

Autoxidative formation is mostly unwanted because it normally gives rise to off-flavours. These are often indicative of deteriorated food and act as a sign of warning against consumption. Chemically, radical reactions with oxygen take place, which produce short chain carboxylic acids, aldehydes and ketones, responsible for rancid notes. These reactions occur especially with polyunsaturated fatty acids as substrates [32, 33].

5.1.3   Flavour Analysis

Analytical chemistry is the basic science to elucidate new flavour compounds. The identification of a new compound is the successful end of an isolation procedure which is schematically shown in Figure 5.4. It is up to the skill of the analyst to choose the right conditions to concentrate, separate and isolate a

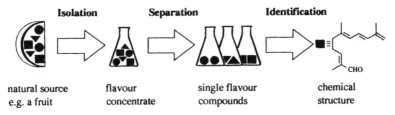

Figure 5.4   The principal steps in flavour analysis.

new substance without its destruction. It can be a fascinating adventure to trace a new compound at the start of an analysis from its aroma until finally the spectroscopic data of an isolated sample allow elucidation of its chemical structure.

The relevance of a new flavour compound can be the novelty of its chemical structure or the importance of its odour note ('impact chemical'). For more detailed descriptions of separation procedures the reader is referred to some recent monographs [34–36]. Further references on instrumental methods and flavour analysis research in food science are also listed [37–39].

A particularly efficient method for flavour research is the 'headspace technique' combined with GLC-MS coupling. In this way it is possible to collect enough volatile materials from a handful of berries or flowers to determine the constituents. It was especially the combination of mass spectrometry (MS) with gas liquid chromatography (GLC) which brought about an enormous increase in sensitivity of detection in the 1970s.

A 'first revolution' in the analysis of volatile compounds had been brought about in the 1950s by the development of GLC. Using this tool it was possible to separate and isolate tiny amounts of substances. It complemented existing procedures because in this technique separation is improved by using very small amounts. (For separation by distillation, the converse is true.) The technical improvements have reached an impressive level during the last 40 years. An apparatus of early days equipped with short, thick columns (e.g. $3\,m \times 5\,mm$) and a thermal conductivity detector had a practical detection limit in the range of milligrams to micrograms (1×10^{-3}–1×10^{-6} g). Its modern successor equipped with a capillary column (e.g. $40\,m \times 0.3\,mm$) and a flame ionisation detector (or a nitrogen or sulphur sensitive detector) detects traces in the range of picograms (1×10^{-12} g).

Another development of the 1980s is high performance liquid chromatography (HPLC) which makes possible separation at normal temperatures of thermally labile compounds in a way that is independent of volatility and polarity. Each method has its own complementary value within the repertory of analytical procedures.

The modern development of equipment and interpretation of results would not have been possible without electronically based data handling and storage. For example, the measured mass spectra of a substance can be compared within a very short time with thousands of reference spectra making possible a crude analysis (~ 90%) within a few days. Flavour analysis is carried out in the research facilities of universities (Departments of Food Science, Nutrition, Agriculture and so on) or in those of the food and flavour industry. Research topics at universities are usually of a more academic and fundamental nature and the results are published in scientific journals, whereas industrial research is often more applied and the results are patented or sometimes temporarily withheld to maintain the 'know-how'. Continuous analytical research activities result in a steady increase of knowledge on constituents in all kinds of foodstuffs. The number of volatile compounds in food as published by TNO [1] amounted to 800 in 1967, tripled to 2400 by 1974, and reached 5500 in 1988 (see Figure 5.55). Thanks to new analytical techniques and instrumental methods there is no end to this development, but efforts to identify new and valuable impact chemicals are becoming greater and greater.

5.1.4 Flavour Manufacture

There are three principal methods to obtain flavour materials: to gain the substance from natural sources by physical methods such as extraction, steam distillation and so on; to produce it by a chemical synthesis or to obtain the material as a reaction flavour. A number of criteria determine which method is applicable in a given situation.

For synthetic production of a compound the following conditions should be fulfilled

(a) The chemical structure of the flavour compound responsible for the desired note must be known. Structural elucidation of the flavour compounds will be the target of flavour analysis, and in many instances this work will have been carried out and published [1]. But in cases of very low concentration of a flavour compound, due to its low threshold value or in cases of a delicate balance of several flavour compounds, the indentification of the real impact chemicals can be very expensive or extremely difficult. Yet this is still the best way to elucidate structures of valuable flavour notes, because on one hand, the pure 'trial and error' approach is even more laborious and on the other hand there is still no generally valid correlation of structure and flavour. Hence the design of a molecule with tailor-made olfactory properties is not yet in sight. There are few examples of compounds obeying such a rule (cf. the molecular theory of sweet taste proposed by Kier [40]) and they are rather the exception in structure-activity relationship in human olfaction [41].

(b) The purchasing cost of the flavour material must not exceed a certain limit. More specifically it is the utilisation cost (i.e. the price per kilogram combined with the used dosage) that is important. In the free market every odour or taste has a price range depending on the available flavour materials. A new substance must be placed favourably in this price structure in order to be considered for introduction.

(c) The legislative situation must be clear. National regulations must be considered. In most countries it is easier to introduce a synthetic flavour compound, if it has been identified in one or more common foods, by the argument that intake of a foodstuff is a form of long term toxicological observation.

(d) Last but not least, the consumer of the flavoured end-product must accept and buy it.

If the above considerations are all in favour of synthetic production, then the synthetic work in its strict sense may start. Synthetic chemistry is a science of its own, and only some outlines are given here. Theoretical knowledge, practical expertise and endurance are essential for a successful synthesis. Normally three phases are carried out, from the laboratory (gram scale preparation) to the pilot plant (kilogram scale) and to the production plant (up to ton scale). Each stage has its own problems and difficulties, but let us assume that the desired flavour compound is finally produced and is now at the disposal of the flavourist.

It is estimated that about 80% of all flavourings sold contain at least one synthetic compound, and this indicates the importance of synthetic flavour ingredients.

5.1.5   Composition and Formulation

The development of powerful techniques like GC, coupled GC-MS and HPLC in the last two decades has led to the identification of hundreds of individual flavour components in food and beverages. The work of flavourists is gradually moving away from the traditional approach of composition by imagination and subjective interpretation of flavour ingredients to the current use of formulation, and in the future flavour design. The blending together of flavour ingredients to create a satisfying flavour still remains a highly sophisticated process based on a long practical experience coupled with a very deep knowledge of the subject. Flavourists and flavour chemists have to collaborate very closely to achieve a fully satisfying flavour, capable of meeting all the required processing conditions.

To meet specific applications, tailor-made flavours may seem to some to be today only a matter of raw material puchasing; actually much work remains to be done to reach this aim. Substances identified in food have to be evaluated, either singly or in a mixture according to their threshold level, toxicologically, tested and correctly synthesized in an economical way. More trace impact substances occurring in natural products have yet to be characterized and identified.

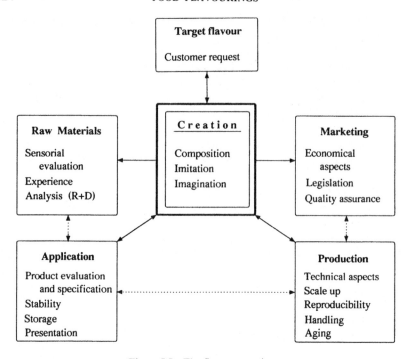

Figure 5.5 The flavour creation.

Until this work is done, the flavourist will continue to solve his problems by using global ingredients as bases in combination with single ingredients, although the computer aided flavour creation (CAFCR) is in view. In the future more and more synthetic flavour ingredients will play a very important role (Figure 5.5).

5.2 Synthetic Flavour Ingredients

5.2.1 *Classification*

The many reasons for use of synthetic flavour ingredients are basically the same as those for the use of flavours themselves (enhancing, replacing, economical price, varying, rounding up, masking etc).

Synthetic flavour ingredients, i.e. nature identical products, cover the whole range of organic substances but no obvious relationship has been established between structure and flavour properties. Some components of similar structure have broadly similar odours but there are many exceptions. The experience of many flavourists 'in the field' is rather discouraging, since the same molecule may be differently perceived at different concentrations.

One of the many possible ways to describe flavour substances or ingredients used in the process of flavour composition is to classify them in groups with parent flavour characteristics as shown in Figure 5.6 (flavour wheel). The same component which belongs to a typical group may be used in quite different flavours. This flavour wheel is not always representative since every flavourist has his own subjective perception of flavour notes, but experience shows that compositions achieved in this way always give similar overall results.

About 2000 single synthetic ingredients are known to be used in flavour compositions, most of them having been identified in natural materials. A small number has not yet been found in nature but recognised as safe and therefore permitted by most countries (e.g. ethyl vanillin, ethyl maltol). Another 3000 substances arising from new raw materials such as process or biotechnological flavours are used as building blocks. As mentioned above, flavour notes are not necessarily bound to specific structures of chemical classes. For flavourists classification according to sensory properties is more important and we discuss the different synthetic flavour ingredients according to their properties as shown in Figure 5.6.

5.2.2 The Flavour Wheel

The flavour wheel is a pictorial illustration of some basic flavour relations. At the centre is the flavour matrix which represents the body of the flavour to be created. It contains all the ingredients needed to support, dilute, enhance and protect the single flavour components, which cannot be applied in a pure state.

Depending on the application, different carriers (solid, liquid, polar, apolar) and enhancers (sweet, savoury, salty) as well as intermediate rounding up compounds (extracts, building blocks) are widely utilised. Around the flavour matrix, segments are arranged according to the individual flavour notes and illustrated by some typical examples. The sequence of the segments is chosen in such a way that the flavour notes are adjacent to kindred characteristics, comparable to the array of the spectral colours in a rainbow. In principle a global flavour perception can be partitioned into a number of 'pure' flavour notes, and rearranged again by composing these notes in the right proportions in an appropriate flavour matrix. Theoretically a correct flavour profile description is simultaneously a recipe for the reconstruction of this flavour. The flavour wheel will give some insights into the most important flavour notes with their possible reactions. This may be useful for flavour chemists but is not intended to replace the experience and creativity of flavourists.

The sections treating the flavour notes from 5.2.3.1 to 5.2.3.16 give a short general description of each flavour, some information about occurrence and origin of typical flavour compounds and examples of their use. Paragraphs and schemes on biogenesis and synthesis of specific examples conclude each flavour segment.

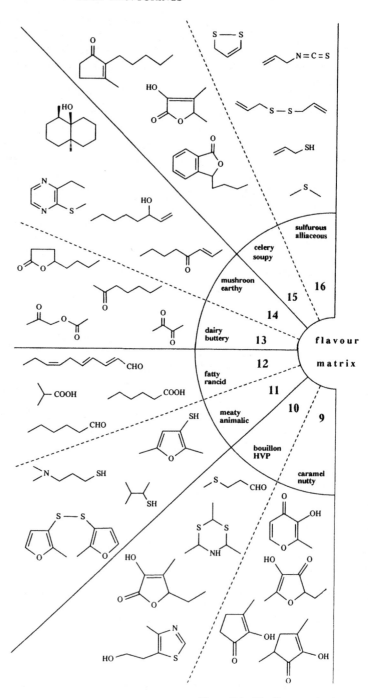

Figure 5.6   The flavour wheel.

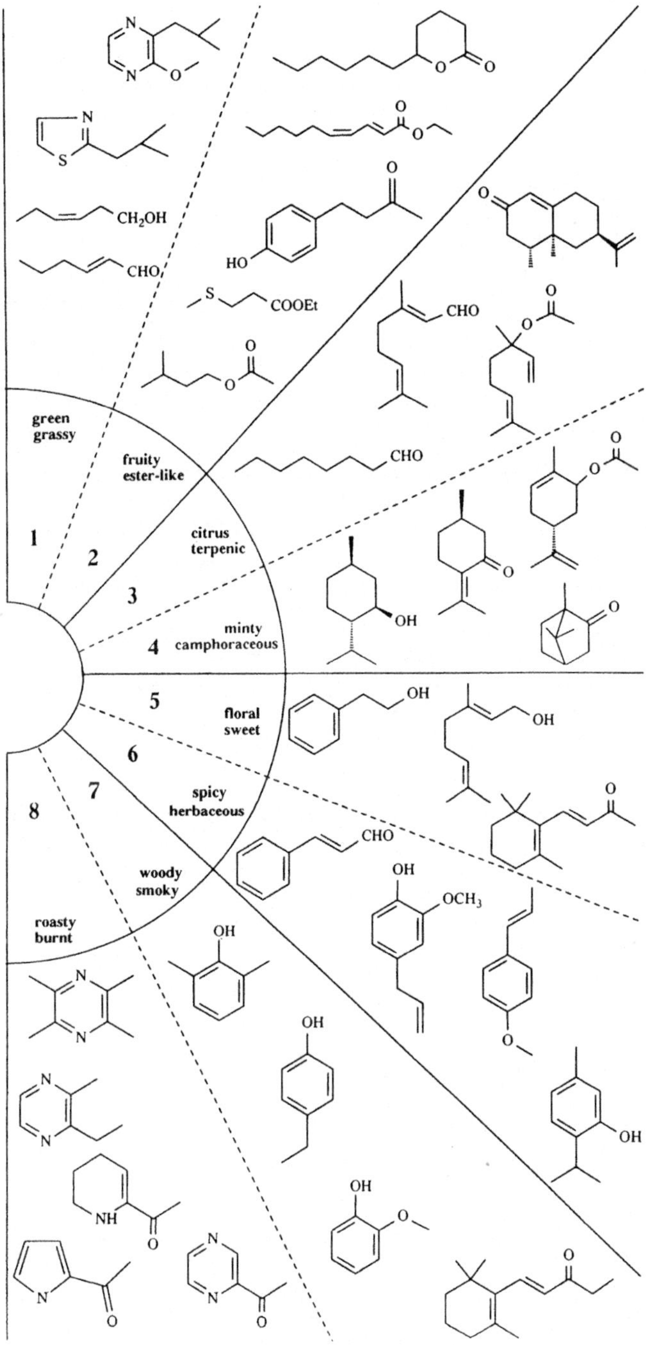
green
grassy
1
fruity
ester-like
2
citrus
terpenic
3
minty
camphoraceous
4
5
floral
sweet
6
spicy
herbaceous
7
woody
smoky
8
roasty
burnt
CH2OH
CHO
COOEt
HO
OH
OCH3

5.2.3   *The Different Flavour Notes*

5.2.3.1   *The green grassy flavour note.*   The green flavour note is described as the odour of freshly cut grass or ground leaves and green plant materials. Typical substances representing this class are short chain unsaturated aldehydes and alcohols such as *trans*-2-hexenal and *cis*-3-hexenol. Esters and heterocycles like alkyl substituted thiazoles and alkoxy pyrazines with very low flavour threshold levels also belong to this group (Figure 5.7).

Figure 5.8 shows the well-known biogenetic pathway for the formation of C_6-unsaturated aldehydes and alcohols [42]. *Cis*-3-Hexenal, *trans*-2-hexenal, *cis*-3-hexenol and *trans*-2-hexenol are very important impact substances occurring in many fruits and vegetables (apple, tomato, grapes, etc.). The enzymatic formation of thiazoles and pyrazines is suggested as originating from amino acids, but their biogenetic pathways have not yet been clarified.

Many other compounds like esters, acids and terpenoids could be placed in this group as 'rounding up' green notes (e.g. hexyl 2-methyl-butyrate, α-pinene). Only *trans*-2-hexenal and *cis*-3-hexenol are commonly used in flavour compounding because of their stability and commercial availability. 2-Isobutyl thiazole is used only in specific notes (e.g. tomatoes). Figure 5.9 shows synthetic routes for the preparation of *cis*-3-hexenol and 4,5-dimethyl-2-pentyl thiazole, respectively. The synthesis of *cis*-3-hexenol (Figure 5.9a) is straightforward, the important step being the selective hydrogenation of the triple bond to the *cis* isomer [43]. Several different preparations are known today. The first manufacturing method introduced in the early 1960s started a 'green period' in the flavour and fragrance industry. The annual world consumption of leaf alcohol is estimated to be 40 tons [44].

The thiazole synthesis follows a procedure of Dubs [45], beginning with the appropriate starting materials as shown in Figure 5.9b. The same procedure

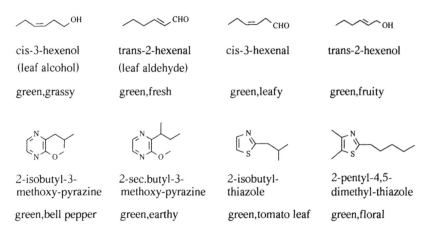

cis-3-hexenol	trans-2-hexenal	cis-3-hexenal	trans-2-hexenol
(leaf alcohol)	(leaf aldehyde)		
green,grassy	green,fresh	green,leafy	green,fruity

| 2-isobutyl-3-methoxy-pyrazine | 2-sec.butyl-3-methoxy-pyrazine | 2-isobutyl-thiazole | 2-pentyl-4,5-dimethyl-thiazole |
| green,bell pepper | green,earthy | green,tomato leaf | green,floral |

Figure 5.7   Examples of green grassy flavour notes.

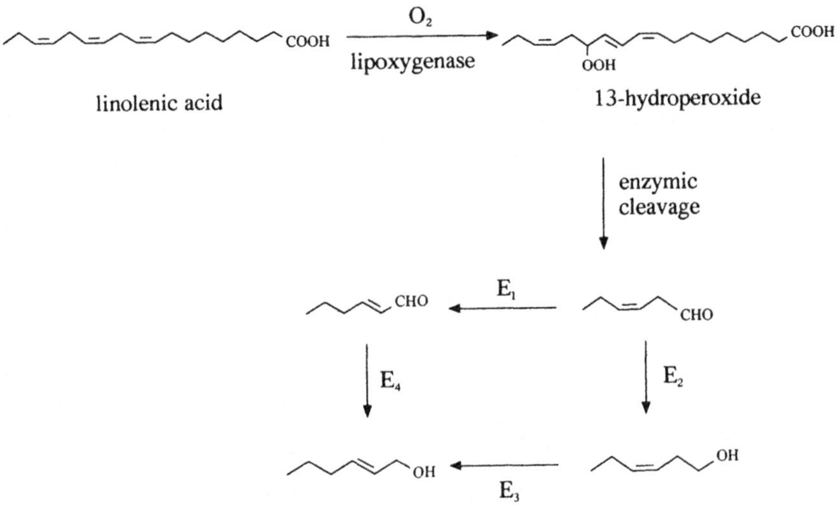

Figure 5.8   Enzymatic formation of C_6-compounds from linolenic acid [42]. $E_1, E_2,...,$ different enzyme systems.

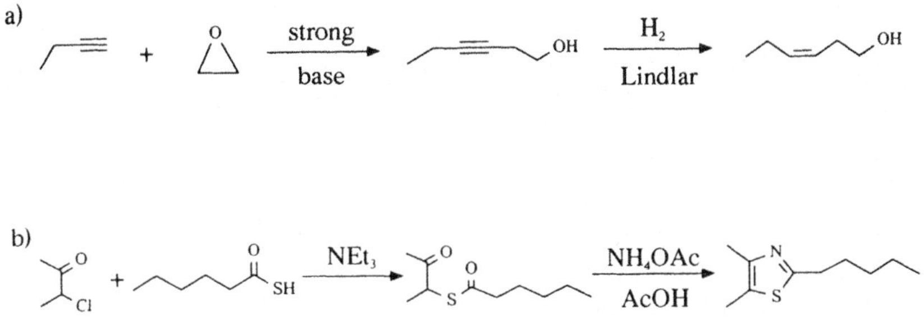

Figure 5.9   Synthesis of (a) *cis*-3-hexenol [44] and (b) 4,5-dimethyl-2-pentythiazole [45].

can be used to prepare other alkyl substituted thiazoles, usually in good yields. Many of these alkyl thiazoles display green flavour notes in numerous 'shades'.

5.2.3.2 *The fruity ester-like flavour note.* The fruity ester-like note is characterised as the sweet odours occurring generally in ripe fruits such as banana, pear, melon, etc. Typical ingredients representing this group are esters and lactones but ketones, ethers and acetals are also involved (Figure 5.10) . Tropical fruits are known to contain sulphur compounds responsible for the typical 'exotic' notes (e.g. the methyl and ethyl ester of 3-methylthiopropionic acid).

Isoamylacetate is responsible for the sweet character in almost all fruity notes. 2,4-Decadienic acid ester is the impact substance found in Bartlett pears

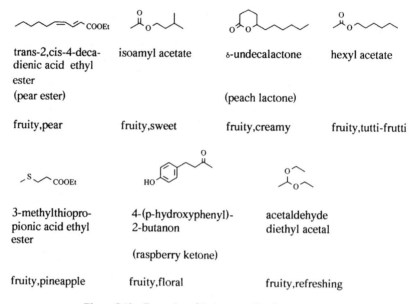

trans-2,cis-4-deca-
dienic acid ethyl
ester
(pear ester)

fruity,pear

isoamyl acetate

fruity,sweet

δ-undecalactone

(peach lactone)

fruity,creamy

hexyl acetate

fruity,tutti-frutti

3-methylthiopro-
pionic acid ethyl
ester

fruity,pineapple

4-(p-hydroxyphenyl)-
2-butanon

(raspberry ketone)

fruity,floral

acetaldehyde
diethyl acetal

fruity,refreshing

Figure 5.10 Examples of fruity ester-like flavour notes.

[46]. 3-Methylthiopropionic acid esters are characteristic for pineapple. Those components are also used with many other flavour notes to give special characters. Additional compounds with non-specific fruity characters but generally described as sweet and fresh (like lactones, aromatic aldehydes, lactones and terpenoids) can also be used to round up fruity notes.

The formation of 3-methylthiopropionic esters can be assumed to start from methionine by Strecker degradation, followed by oxidation and esterification. Figure 5.11 shows the formation of the impact substance *trans-2,cis-4-*decadienic acid ethyl ester found in pear, apple and grape. Jennings and Tressl

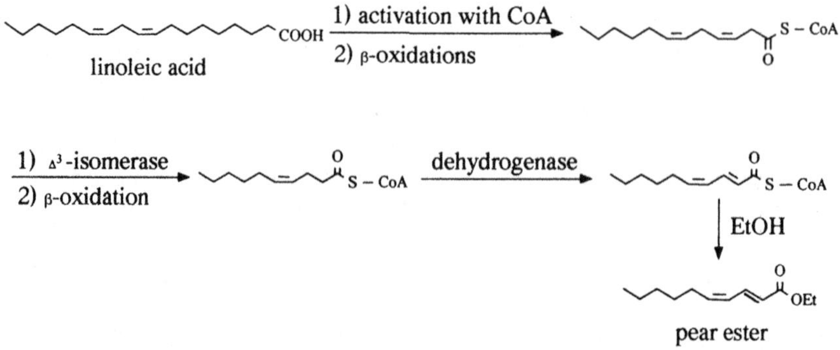

linoleic acid → 1) activation with CoA 2) β-oxidations → S–CoA

1) Δ³-isomerase 2) β-oxidation → S–CoA → dehydrogenase → S–CoA

EtOH

pear ester

Figure 5.11 Formation of ethyl *trans-2,cis-4* decadienoate (pear ester) [46].

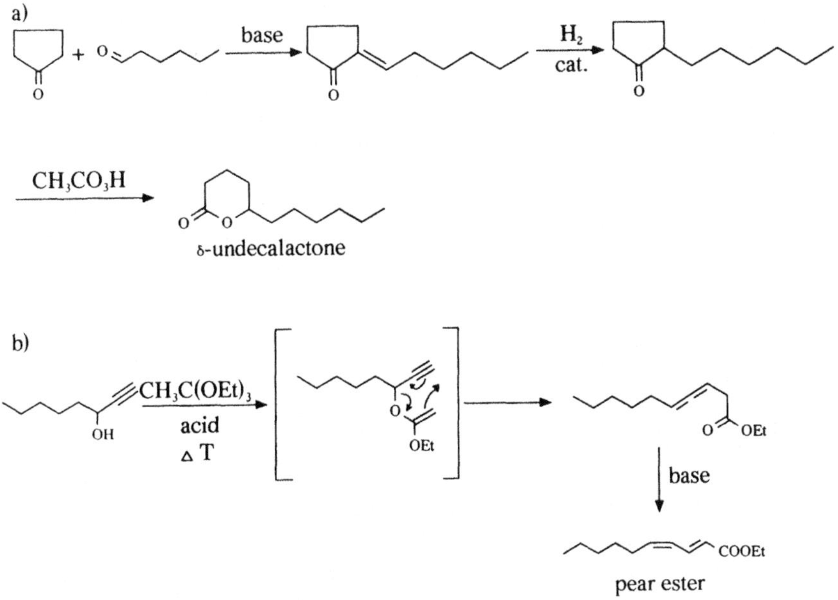

Figure 5.12 Synthesis of (a) δ-undecalactone [47] and (b) ethyl *trans*-2,*cis*-4-decadienoate (pear ester) [48].

proposed an interesting scheme to explain their analytical findings during post-harvest ripening of Bartlett pears. Linoleic acid is degraded by several β-oxidations, each degrading cycle shortening the chain by two carbons [46]. The degradation can be explained using the same enzymes of the fatty acid metabolism as in the formation, each enzyme-catalysed reaction can occur in both directions.

Figure 5.12a demonstrates, by example with δ-undecalactone, a simple and efficient synthesis of 6-membered lactones with a variable chain in the 5-position [47]. In Figure 5.12b, an extension of a Claisen rearrangement is applied to prepare the 'pear ester'. The reaction conditions and the kind of base in the isomerisation step are crucial for a good yield of the wanted 2-*trans*, 4-*cis* double bond system [48]. The other possible isomers are less desired, due to more fatty and green flavour notes.

5.2.3.3 *The citrus terpenic flavour note.* The citrus-like flavour note is characterised as the typical note occurring in citrus fruits and plants (citrus, lemon, orange, grapefruit), but also certain terpenoids can occur in this group. One of the most important components with a strong impact odour is citral (a mixture of geranial and neral), which can be either isolated from natural raw materials (lemon grass) or synthesised (see Figure 5.13). Another terpenic component with a rather more bitter character is nootkatone, which is an

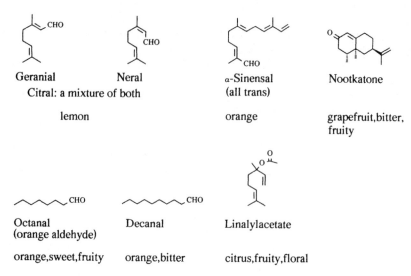

CHO
CHO
CHO
O
Geranial
Neral
Citral: a mixture of both
α-Sinensal
(all trans)
Nootkatone
lemon
orange
grapefruit,bitter,
fruity
O
O
CHO
CHO
Octanal
(orange aldehyde)
Decanal
Linalylacetate
orange,sweet,fruity
orange,bitter
citrus,fruity,floral

Figure 5.13 Examples of citrus terpenic flavour notes.

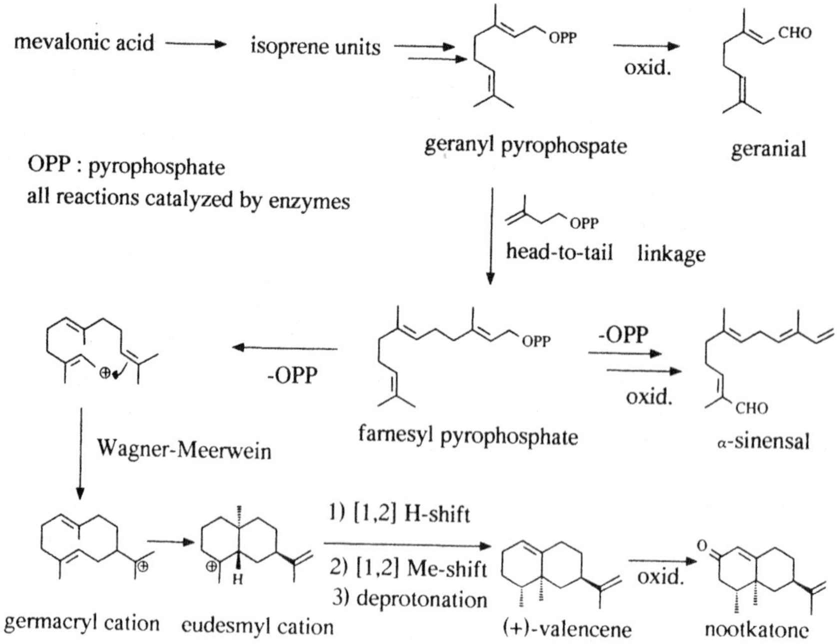

mevalonic acid
isoprene units
OPP
oxid.
CHO
geranyl pyrophospate
geranial
OPP : pyrophosphate
all reactions catalyzed by enzymes
OPP
head-to-tail linkage
-OPP
OPP
-OPP
oxid.
CHO
farnesyl pyrophosphate
α-sinensal
Wagner-Meerwein
1) [1,2] H-shift
2) [1,2] Me-shift
3) deprotonation
H
oxid.
O
germacryl cation
eudesmyl cation
(+)-valencene
nootkatone

Figure 5.14 Biogenesis of geranial, α-sinensal and nootkatone [20]. OPP, pyrophosphate; all reactions catalysed by enzymes.

impact substance found in grapefruit. Simple aliphatic aldehydes of medium chain length (octanal, decanal), but also sinensal, an unsaturated C_{15}-aldehyde, and some esters of monoterpenic alcohols (linalyl acetate) are used to round up this note.

In Figure 5.14, the biogenetic pathways of three representatives of terpenoid structure and their common precursors, i.e. mevalonic acid and 'isoprene units' are shown. The formation of geranial can be easily explained by an oxidation of geranyl pyrophosphate, which in turn is the regular head-to-tail linkage product of two isoprene units as outlined in Figure 5.2. α-Sinensal is a regular sesquiterpene formed from farnesyl pyrophosphate (a further head-to-tail elongation of geranyl pyrophosphate), though the oxidation took place at the 'head' of the molecule. Whereas geranial and α-sinsensal are examples of regular acyclic terpenes, nootkatone must have been formed by two cyclisation steps and a [1, 2]-methyl shift, which ultimately ends up in an irregular isoprene pattern.

Besides natural citral which is frequently preferred for its harmony, standardised synthetic citral is generally used when large amounts and low costs are needed. Citral is an important synthetic ingredient with a total consumption of about 100 tons in 1985 for flavours, and an annual production of several thousand tons in the United States [44]. Figure 5.15a shows a

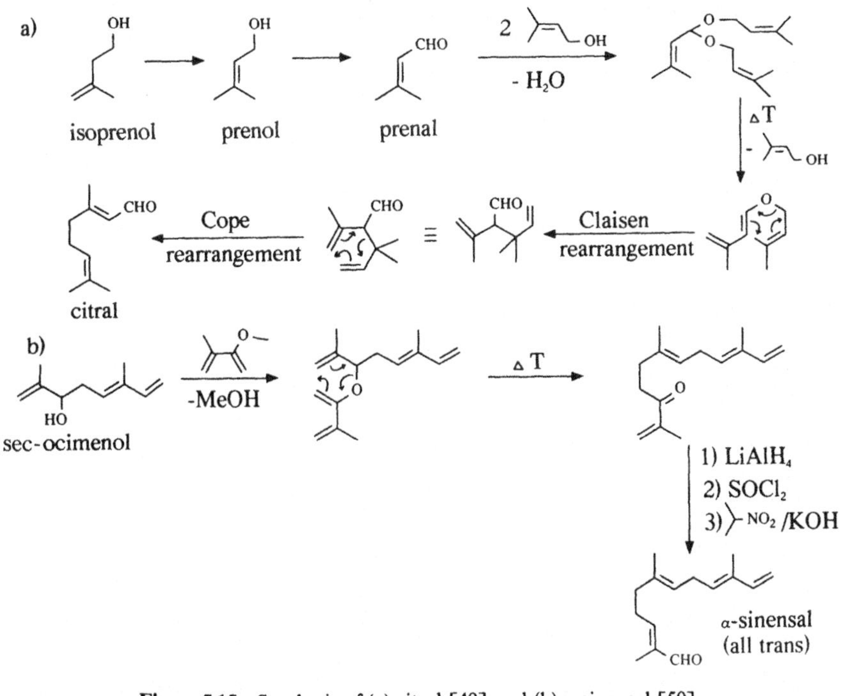

Figure 5.15   Synthesis of (a) citral [49] and (b) α-sinensal [50].

technical synthesis of citral [49] and one of the published syntheses of all-*trans*-α-sinensal (b) is shown in Figure 5.15b [50, 50a]. The citral synthesis is very elegant and short, because the necessary temperature for the elimination of prenol triggers the two consecutive rearrangements in excellent yields. Because of the importance of citral, other synthetic procedures of industrial scale have been patented [51, 52].

In comparing Figures 5.14 and 5.15, it is remarkable to see the similarity between the biogenetic pathway and the chemical synthesis; in both cases an unsaturated C_5-unit seems to fit best. This fact can be observed frequently and relies on the inherent chemical reactivity of a compound.

5.2.3.4 *The minty camphoraceous flavour note.*

The minty note is described as the impact of peppermint having a sweet, fresh and cooling sensation. The main compounds are *l*-menthol, pulegone, *l*-carvylacetate, *l*-carvone and camphor, but borneol, eucalyptol (= cineol), and fenchone also belong traditionally to this odour family (see Figure 5.16). The minty notes are generally used as refreshing top notes with cooling effects in special flavours and applications (chewing gum, toiletries and beverages).

In this group *l*-menthol is economically the most important synthetic ingredient, the main quantity being used in non-food applications. Most of these minty camphoraceous compounds are oxygenated monoterpenes and are formed according to the general outline of Figure 5.2. As shown in Figure 5.17, the formation of menthol can be explained by cyclisation of linalyl cation followed by adjustment of the functional groups (introduction of

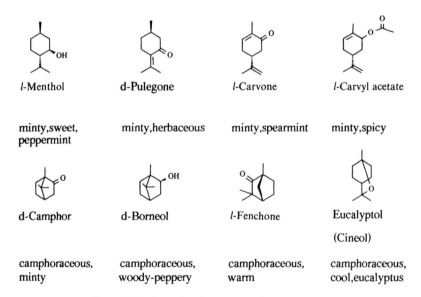

l-Menthol	d-Pulegone	*l*-Carvone	*l*-Carvyl acetate
minty, sweet, peppermint	minty, herbaceous	minty, spearmint	minty, spicy

d-Camphor	d-Borneol	*l*-Fenchone	Eucalyptol (Cineol)
camphoraceous, minty	camphoraceous, woody-peppery	camphoraceous, warm	camphoraceous, cool, eucalyptus

Figure 5.16 Examples of minty camphoraceous notes.

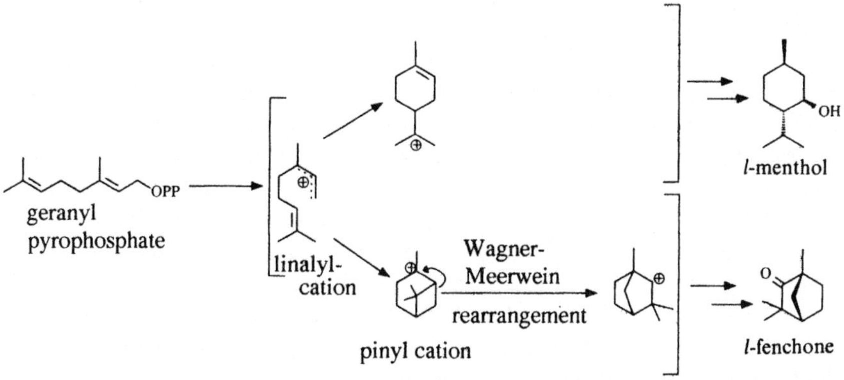

Figure 5.17   Biogenesis of menthol and fenchone.

hydroxyl, hydrogenation of double bonds). The formation of fenchone is thought to proceed as an enzyme bound cation, since no free intermediates are involved in this conversion catalysed by the enzyme *l*-endo-fenchol-synthetase (isolated from fennel) [53]. It is, however, evident that a re-arrangement of the carbon skeleton must occur, because fenchone belongs to the irregular monoterpenes. The stereochemical outcome of the reaction sequence (only one enantiomer is formed) is in accordance with the enzymatic control of each step from geranyl pyrophosphate to the end-product.

In most applications *l*-menthol is the preferred isomer of menthyl alcohols. In Figure 5.18 an industrial systhesis is shown, which produces specifically the desired isomer. The crucial step is the enantioselective isomerisation of geranyl diethyl amine with an homogeneous asymmetric rhodium catalyst. Another industrial preparation includes the resolution of a mixture of menthol,

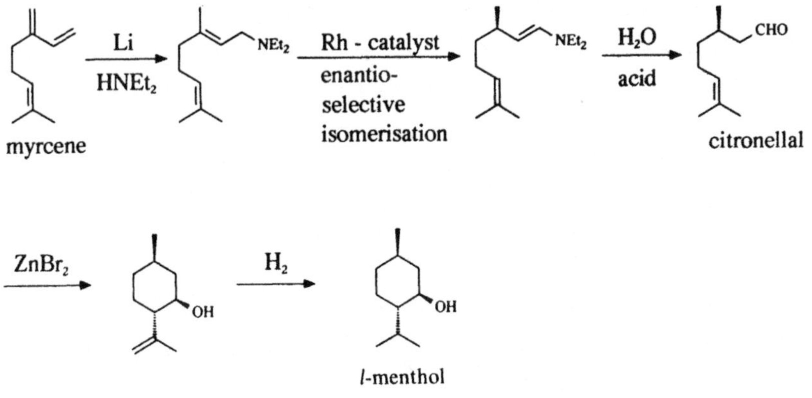

Figure 5.18   Synthesis of *l*-menthol [54].

neomenthol and isomenthol [55]. Due to these and other successful syntheses *l*-menthol is a low price item, available in large quantities.

5.2.3.5 The floral sweet flavour notes.   The floral note can be defined as the odours emitted by flowers and contains sweet, green, fruity and herbaceous characters. There are no typical impact ingredients, but a range of single flavour substances belonging to different chemical classes (Figure 5.19). Phenylethanol, geraniol, *β*-ionone and some esters (benzyl acetate, linalyl acetate) are important compounds in this group. All these compounds originate from plant materials. At higher concentrations many floral notes are accompanied by an unwanted perfume-like impression.

The formation of *β*-ionone (Figure 5.20a) includes the formation and degradation of *β*-carotene. The formation is yet another example of terpene biogenesis, two farnesyl pyrophosphate moieties being fused, tail-to-tail, to give the C_{30}-compound squalene. Further enzymatic steps form *β*-carotene, which is oxidatively split into smaller odoriferous fragments like *β*-ionone and other substances later in the ripening process [56].

The reasons why a plant generates and metabolises higher terpenes like *β*-carotene is not well understood. Oxidative degradation can happen as part of an enzymatic ripening process or as a post-harvest (photo-oxidative or auto-oxidative) event.

The formation of 2-phenylethanol (Figure 5.20b) is an example of a Strecker degradation of the amino acid phenylalanine (cf. Figure 5.1b). Tressl et al. showed that in banana tissue slices, the radioactive label of added L-

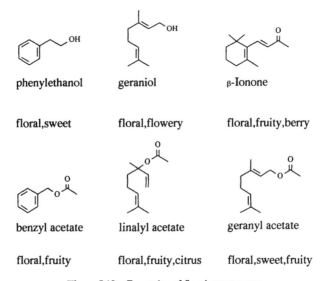

phenylethanol	geraniol	β-Ionone
floral,sweet	floral,flowery	floral,fruity,berry

benzyl acetate	linalyl acetate	geranyl acetate
floral,fruity	floral,fruity,citrus	floral,sweet,fruity

Figure 5.19 Examples of floral sweet notes.

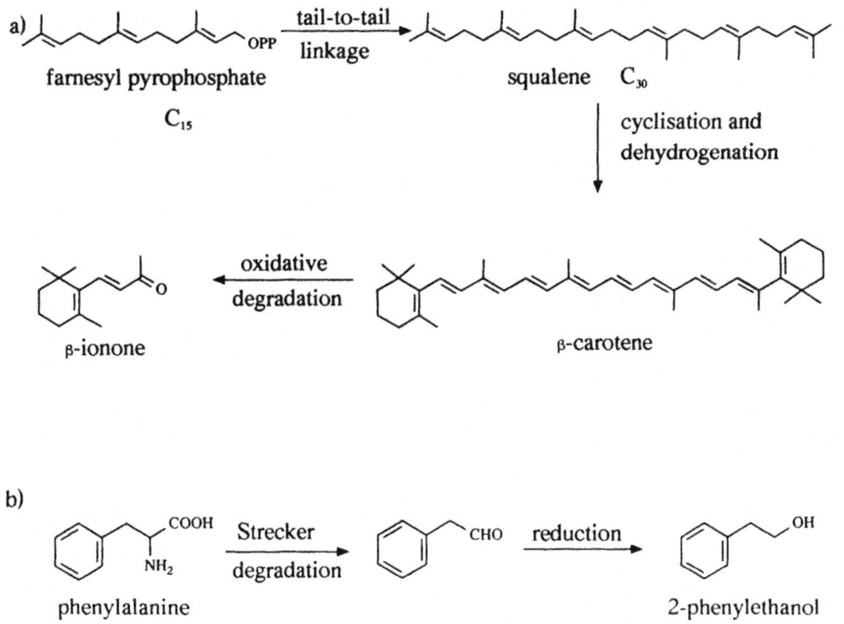
a)
OPP
tail-to-tail
linkage
farnesyl pyrophosphate
C15
squalene C30
cyclisation and
dehydrogenation
oxidative
degradation
O
β-ionone
β-carotene
b)
COOH
NH2
Strecker
degradation
CHO
reduction
OH
phenylalanine
2-phenylethanol

Figure 5.20 (a) Biosynthesis of β-carotene and degradation to β-ionone [56]. (b) Generation of 2-phenylethanol [57].

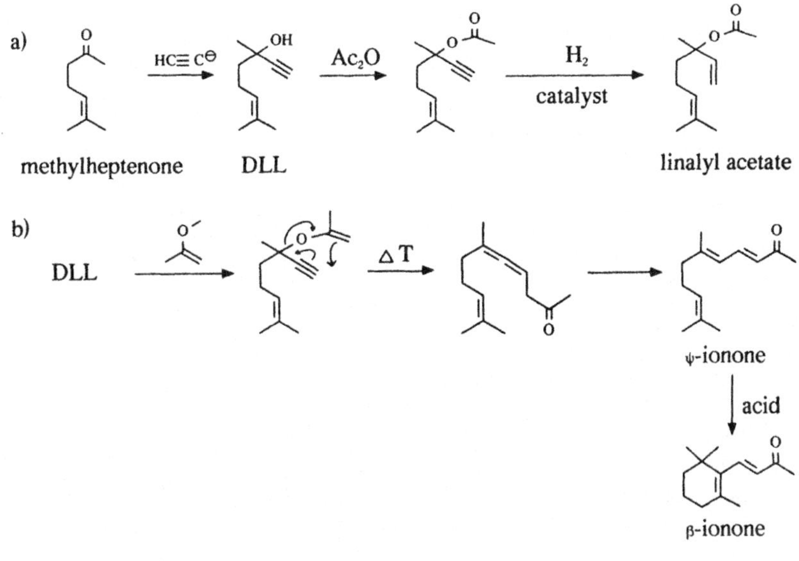
a)
O
HC≡C⊖
OH
Ac2O
O
H2
catalyst
methylheptenone
DLL
linalyl acetate
b)
DLL
O
ΔT
ψ-ionone
acid
β-ionone

Figure 5.21 Synthesis of (a) linalyl acetate [58, 58a] and (b) β-ionone [59].

[2-¹⁴C]phenylalanine could be mainly found, after 5 h of incubation, in 2-phenylethanol [57].

The examples of Figure 5.19 are the largest volume items in the flavour and fragrance industry. The consumption of linalyl acetate in the U.S. market is estimated to be about 500 tons annually [58].

Figure 5.21 shows synthetic procedures for linalyl acetate and β-ionone, based on methyheptenone which is another very large volume commodity. In the acetylene process, dehydrolinalool (DLL) is first produced, and then acetylated and partially hydrogenated (Figure 5.21a) [58, 58a]. The elongation of DLL with a C_3-unit makes use of a modified Claisen rearrangement [59]. Isomerisation of the allenic systems and acid catalysed cyclisation conclude the synthesis of β-ionone (Figure 5.21b).

Besides these purely synthetic approaches other manufacturers start from natural sources, e.g. pinene is transformed into linalyl acetate and geranyl acetate [60].

5.2.3.6   *The spicy herbaceous flavour note.*   The spicy herbaceous flavour note is common to herbs and spices with all their nuances. These flavourings are certainly the oldest food ingredients used by mankind. Spices were especially desired not only as appetisers but also as food preservatives. Aromatic aldehydes, alcohols and phenolic derivatives are typical constituents with their strong flavour and physiological (e.g. bacterio-static) effects. Many are impact character chemicals: anethole (anise), cinnamaldehyde (cinnamon),

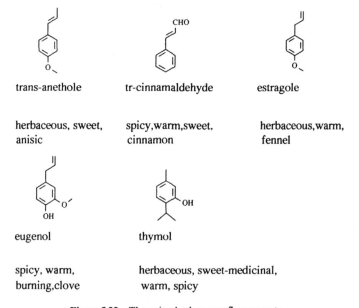

trans-anethole   tr-cinnamaldehyde   estragole

herbaceous, sweet,   spicy,warm,sweet.   herbaceous,warm,
anisic   cinnamon   fennel

eugenol   thymol

spicy, warm,   herbaceous, sweet-medicinal,
burning,clove   warm, spicy

Figure 5.22   The spicy herbaceous flavour note.

estragole (estragon), eugenol (clove), *d*-carvone (dill), thymol (thyme), etc. (Figure 5.22). Due to their strength many of these compounds are used only in small quantities. Often the spices or extracts of them are used. The range of application is very broad and includes bakery goods and alcoholic beverages as well as toothpaste and chewing gum and even fragrances of oriental type. The largest item used is cinnamaldehyde, which is also synthetically produced in considerable amounts.

The formation of cinnamaldehyde, anethole, estragole and eugenol is related to the phenylpropanoid metabolism, i.e. the transformation of phenylalanine to cinnamic acid and its hydroxylated derivatives (Figure 5.23).

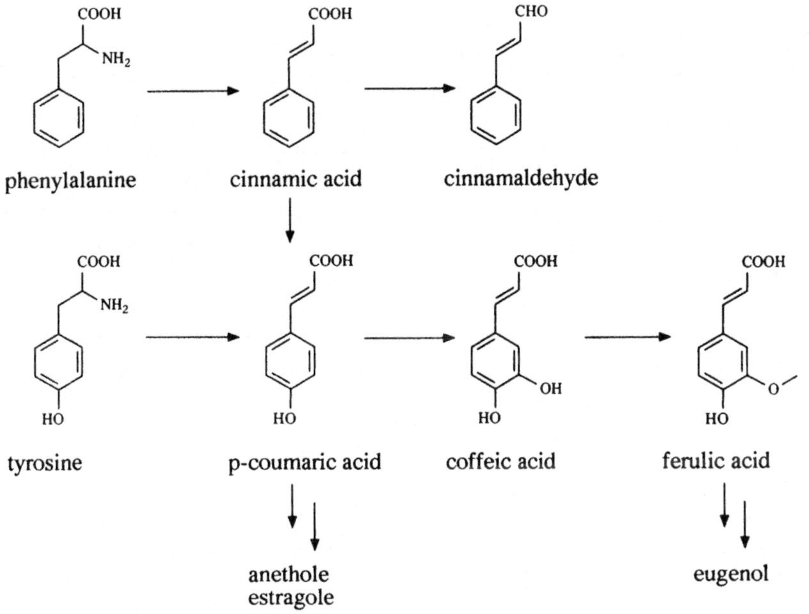

Figure 5.23   Biogenesis of cinnamaldehyde and related phenylpropanoid structures [61].

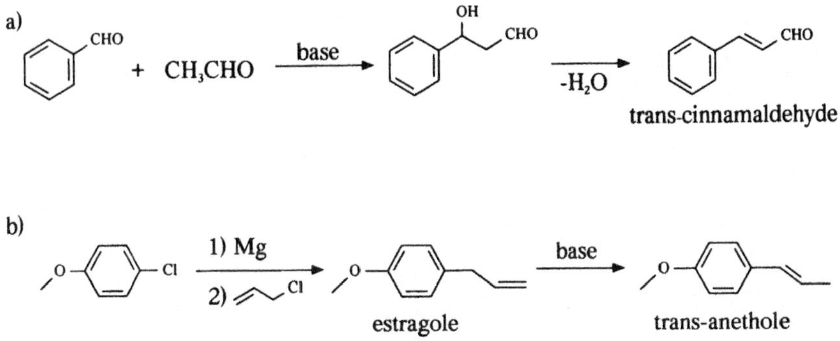

Figure 5.24   Synthesis of (a) cinnamaldehyde [62] and (b) estragole and anethole (unpublished).

The formation of phenylalanine is linked with the carbohydrate metabolic pathway through shikimic acid. In higher plants, the phenylpropanoid pathways are especially important and give rise to a number of compounds in relatively high concentrations [61].

Thymol has its origin in a monoterpenic *p*-menthane derivative, e.g. pulegone. Synthetic cinnamaldehyde is made by an aldol condensation of benzaldehyde with acetaldehyde, a procedure which goes back well into the last century (Figure 5.24a) [62].

Estragole can be synthesised from *p*-chloroanisole by a Grignard reaction with allyl chloride. Base-catalysed isomerisation of the double bond gives *trans*-anethole (Figure 5.24b) (unpublished results).

5.2.3.7 The woody smoky flavour note. Woody smoky flavour notes are characterised by such compounds as substituted phenols (guaiacol etc.), methylated ionone derivatives (methylionone), and, exceptionally, by some aldehydes (*trans*-2-nonenal in very low concentration), which have respectively warm, woody, sweet and smoky odours (Figure 5.25).

Methylionones are examples of synthetic ingredients that have not been found in food although they are structurally very similar to the carotene degradation products, the ionones. The methylionones are regarded as safe [2] and may be used in food flavours in many countries.

The woody and smoky notes are generally considered not to belong to the native flavour compounds, rather they are formed during storage or heat treatment. These compounds can only be used in such combination with

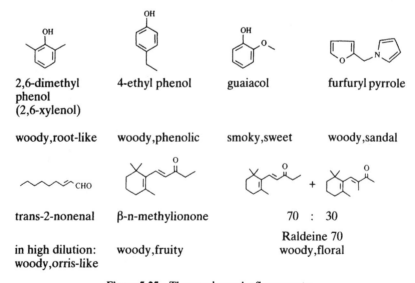

Figure 5.25   The woody smoky flavour note.

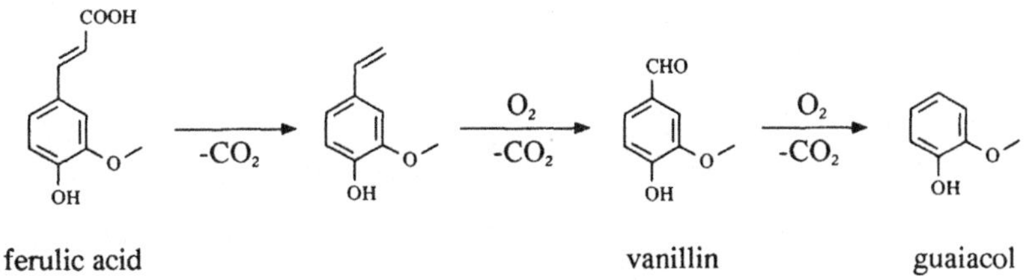

ferulic acid   vanillin   guaiacol

Figure 5.26   Formation of guaiacol by oxidative degradation [63].

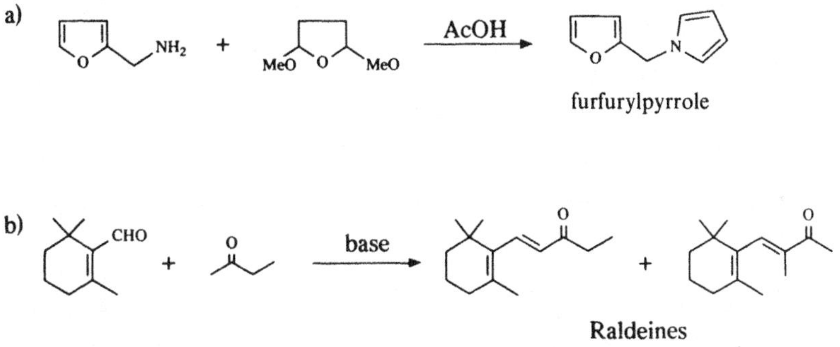

a)

furfurylpyrrole

b)

Raldeines

Figure 5.27   Synthesis of (a) furfurylpyrrole [64] and (b) methylionones [65].

flavour notes where an improvement of mouthfeel and body is needed. In higher concentrations they are perceived as off-flavours.

While the phenols are generally available from processed flavour preparations (e.g. by dry sawdust distillation), ionone derivatives are individually synthesised. The formation of the phenolic derivatives can be explained by thermal degradation of lignin (cf. smoking processes), or by fermentative oxidations (cf. curing of vanilla beans) of phenlpropanoid precursors (Figure 5.26, see also Figure 5.23) [63]. The synthesis of furfurylpyrrole starts from furfuryl amine and 2,5-dimethoxy-tetrahydrofuran in a widely applicable procedure (Figure 5.27a) [64]. Condensation of cyclocitral with butanone gives a mixture of methylionones. The olfactory value depends on the composition of the mixture, which can be adjusted by variation of basic reaction conditions (Figure 5.27b) [65]. The synthesis of *trans*-2-nonenal is shown in Figure 5.42.

5.2.3.8   *The roasty burnt flavour note.*   The roasty burnt flavour note is generally associated with the term 'pyrazine', which is broadly true since pyrazines are able to cover the whole range of this flavour type. They are a predominant chemical class in roasted products. Different substitution by alkyl, acyl or alkoxy and combinations thereof induces a great diversity of flavour impressions; so burnt, roasted, green, earthy and musty notes can

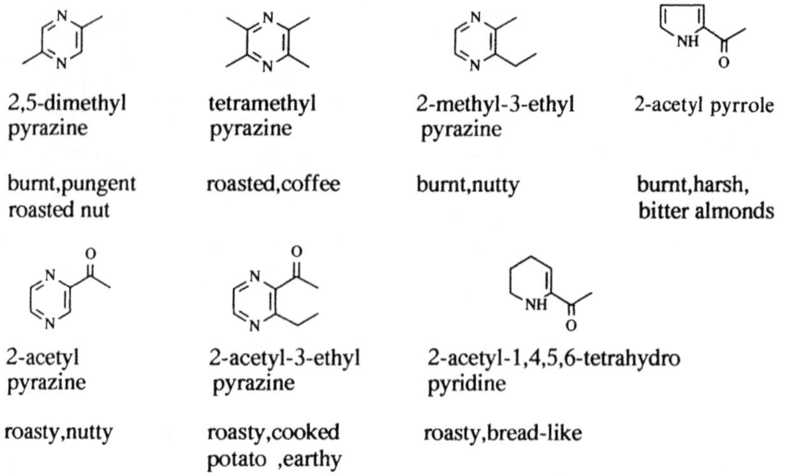
2,5-dimethyl pyrazine
burnt,pungent roasted nut
tetramethyl pyrazine
roasted,coffee
2-methyl-3-ethyl pyrazine
burnt,nutty
2-acetyl pyrrole
burnt,harsh, bitter almonds
2-acetyl pyrazine
roasty,nutty
2-acetyl-3-ethyl pyrazine
roasty,cooked potato ,earthy
2-acetyl-1,4,5,6-tetrahydro pyridine
roasty,bread-like

Figure 5.28 Examples of roasty burnt flavour notes.

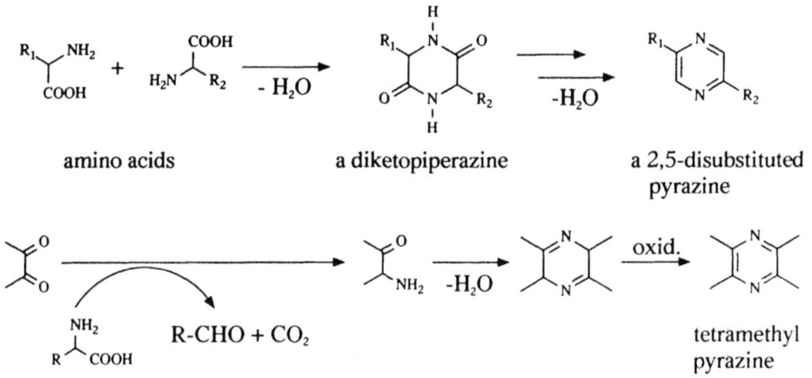
- H2O
-H2O
amino acids
a diketopiperazine
a 2,5-disubstituted pyrazine
oxid.
R-CHO + CO2
tetramethyl pyrazine

Figure 5.29 Formation of pyrazines [65a].

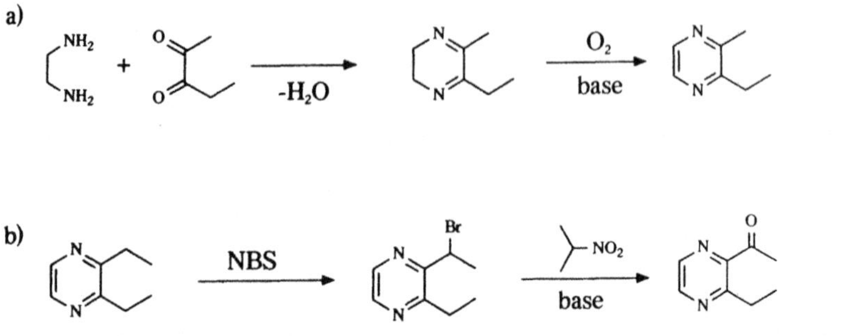
a)
-H2O
O2
base
b)
NBS
base

Figure 5.30 Synthesis of (a) 2-methy-3-ethyl pyrazine [70] and (b) 2-acetyl-3-ethyl pyrazine [71]. NBS, *N*-bromosuccinimide.

result. Within this group of the roasty burnt flavour notes alkyl and acetyl substituted pyrazines are the most important (Figure 5.28).

Several mechanisms and precursors have been proposed to explain the multitude of pyrazines formed in heat treated food (Figure 5.29) [65a]. Condensation of two amino acids leads to a 2,5-diketopiperazine, which is further dehydrated to a pyrazine. Another condensation starts from α-aminoketones, a 'side product' of the Strecker degradation of amino acids (cf. Figure 5.1b). A cyclodehydration reaction of two 1,2-aminoalcohol derivatives (e.g. threonine) was also proposed to form pyrazines.

The predominant role of pyrazines as flavour chemicals is reflected by numerous publications or patents on analytical findings [66] and synthesis [67–69]. Two examples of synthetic procedures for alkyl and acetyl substituted pyrazines are shown in Figure 5.30 [70, 71].

5.2.3.9 The caramel nutty flavour note. The caramel nutty flavour note is mostly described by heat processed sugar-containing food products. The slightly bitter and burnt odour of roasted nuts represents the complement of this flavour type. Beside corylone, maltol, furonol, etc. (see Figure 5.31) a special range of components like vanillin, ethylvanillin, benzaldehyde, phenylacetic acid, cinnamic alcohol, dehydrocoumarin, trimethylpyrazine also belong to this group. The examples of Figure 5.31 are characterised by a common structural unit, namely a cyclic enolone system, and also by a common flavour profile. These cyclic enolones are most valuable compounds for their strong impact character and for their distinct flavour enhancing effect. These

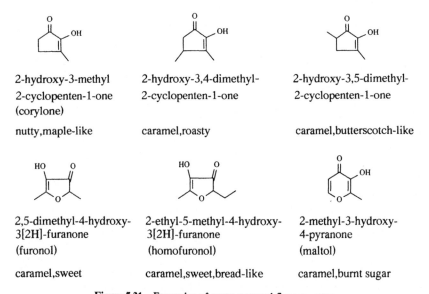

2-hydroxy-3-methyl
2-cyclopenten-1-one
(corylone)

nutty, maple-like

2-hydroxy-3,4-dimethyl-
2-cyclopenten-1-one

caramel, roasty

2-hydroxy-3,5-dimethyl-
2-cyclopenten-1-one

caramel, butterscotch-like

2,5-dimethyl-4-hydroxy-
3[2H]-furanone
(furonol)

caramel, sweet

2-ethyl-5-methyl-4-hydroxy-
3[2H]-furanone
(homofuronol)

caramel, sweet, bread-like

2-methyl-3-hydroxy-
4-pyranone
(maltol)

caramel, burnt sugar

Figure 5.31 Examples of nutty caramel flavour notes.

compounds can be applied in all sorts of caramel flavours and also in roasted and fruit applications.

A comparison of the structures of 18 cyclic enolones and their threshold values confirms an interesting structure-activity relationship. A higher substitution pattern, i.e. more methyl groups or ethyl instead of methyl, lowers the threshold value significantly [72].

The formation of corylone, maltol or furonol occurs during heat treatment in a large range of foodstuffs containing carbohydrates. In Figure 5.32, the anticipated formation of furonol from rhamnose is outlined [73]. This example is a somewhat special case of a Maillard reaction, because the amino acid is replaced by piperidine and acetic acid, and the yield of the end-product is, at around 20%, exceptionally high.

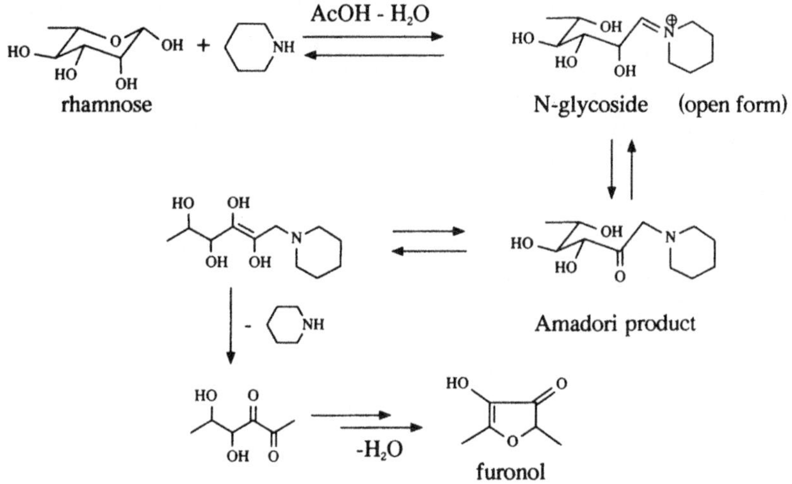

Figure 5.32   Formation of furonol from rhamnose [73].

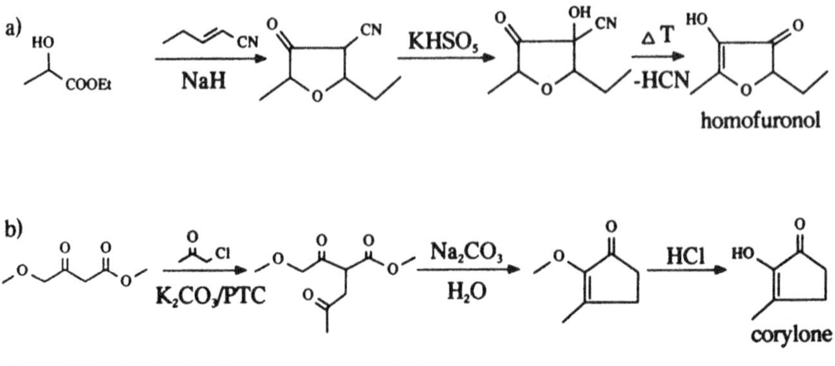

Figure 5.33   Synthesis of (a) homofuronol [74] and (b) corylone [75]. PTC, phase transfer catalysis.

The synthetic routes for homofuronol and corylone as shown in Figure 5.33 are industrially feasible processes [74, 75]. The synthetic concept for homo-furonol allows easy variation of the starting materials, so that with the same reaction sequence, other homologues can be prepared. The synthesis of corylone starts from bulk chemicals using phase transfer catalysis (PTC) conditions. Use of differently alkylated chloroacetones gives rise to higher alkylated corylones (see Figure 5.31 for examples).

5.2.3.10   *The bouillon HVP flavour note.*   The bouillon HVP flavour is not a distinct flavour note based on typical representants, but described as a rather diffuse warm, salty and spicy sensation, associated with enhanced meat extracts. Single flavour ingredients are less important than the appropriate global reaction flavours, which are responsible for the basic taste perception (so-called 'body'). Nevertheless some components are helpful for enhancing and topping (cf. Figure 5.34).

As top notes, 5-methyl-4-mercapto-tetrahydrofuran-3-one and 2-methyl-tetrahydrothiophen-3-thiol are described [76], two compounds which are chemically related to sulphurol degradation products (see also Figure 5.38). The examples of Figure 5.34 exhibit different chemical classes used in this flavour note with one representative of each. 5-Ethyl-4-methyl-3-hydroxy-2-[5H]-furanone ('emoxyfurone') was originally found in HVP. Its formation from threonine can be explained as shown in Figure 5.35a [77]. The odour of this hydroxybutenolide is reminiscent of lovage, where its precursor α-

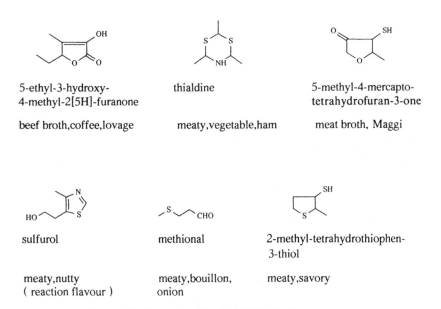

5-ethyl-3-hydroxy-
4-methyl-2[5H]-furanone

beef broth,coffee,lovage

thialdine

meaty,vegetable,ham

5-methyl-4-mercapto-
tetrahydrofuran-3-one

meat broth, Maggi

sulfurol

meaty,nutty
(reaction flavour)

methional

meaty,bouillon,
onion

2-methyl-tetrahydrothiophen-
3-thiol

meaty,savory

Figure 5.34   Examples of bouillon HVP-like flavour notes.

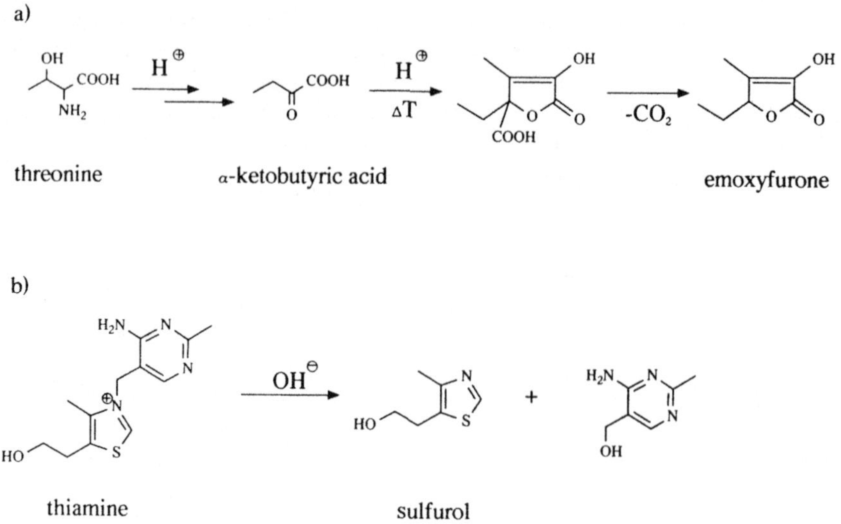

a)

threonine   α-ketobutyric acid   emoxyfurone

b)

thiamine   sulfurol

Figure 5.35 Formation of (a) hydroxybutenolide in HVP [77] and (b) sulfurol from thiamine [78].

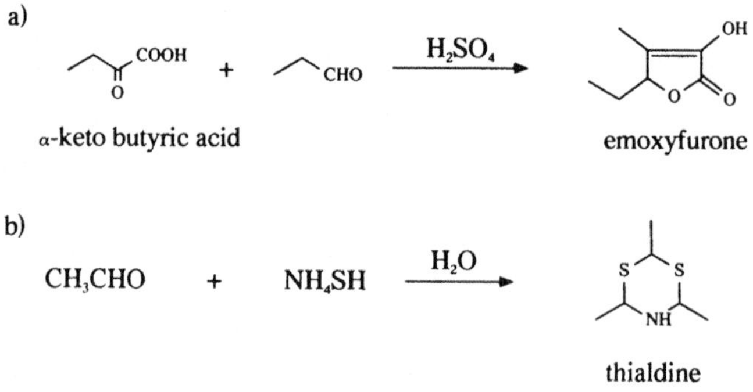

a)

α-keto butyric acid   emoxyfurone

b)

CH_3CHO + NH_4SH $\xrightarrow{H_2O}$

thialdine

Figure 5.36 Synthesis of (a) emoxyfurone [79] and (b) thialdine [80].

ketobutyric acid was identified. 4-Methyl-5-hydroxyethyl thiazole (sulphurol) is described as a thiamine degradation product (Figure 5.35b) [78].

Sulphurol is an example of a curious observation which often irritates flavourists and flavour chemists. Whereas strictly pure sulphurol is a disappointingly weak compound, 'aged' samples display a rich and strong odour of beef broth. Even today the 'impurities' responsible for the odoriferous part cannot be instrumentally detected.

Figure 5.36 shows practical syntheses of emoxyfurone [79] and thialdine [80].

5.2.3.11 *The meaty animalic flavour note* (see Chapter 10). The meaty animalic flavour note is one of the most complex to be described. Roast beef meat flavour differs, for example, from that of barbecued or simply boiled meat. Crude uncooked (bloody animalic) meat is poorly flavoured and reminiscent of a salty, amine taste. The amine flavour note can be evoked by low concentrations of *n*-butyl amine or piperidine. The desired flavour compounds are generated only during processing at the beginning of the Maillard reaction. Sulphur containing components (mercaptans, thiazoles, thiophenes, etc.) as well as nitrogen heterocycles (pyrazines, pyrroles, pyridines, oxazoles, etc.) predominate the flavour effect. Figure 5.37 shows some flavour components.

Depending on the kind and content of fat incorporated or used, a large number of aliphatic carbonyls (aldehydes and ketones) develop and influence

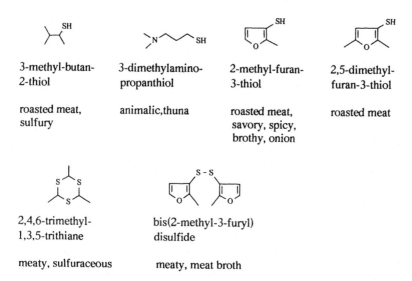

3-methyl-butan-
2-thiol

roasted meat,
sulfury

3-dimethylamino-
propanthiol

animalic,thuna

2-methyl-furan-
3-thiol

roasted meat,
savory, spicy,
brothy, onion

2,5-dimethyl-
furan-3-thiol

roasted meat

2,4,6-trimethyl-
1,3,5-trithiane

meaty, sulfuraceous

bis(2-methyl-3-furyl)
disulfide

meaty, meat broth

Figure 5.37 Examples of meaty animalic flavour notes.

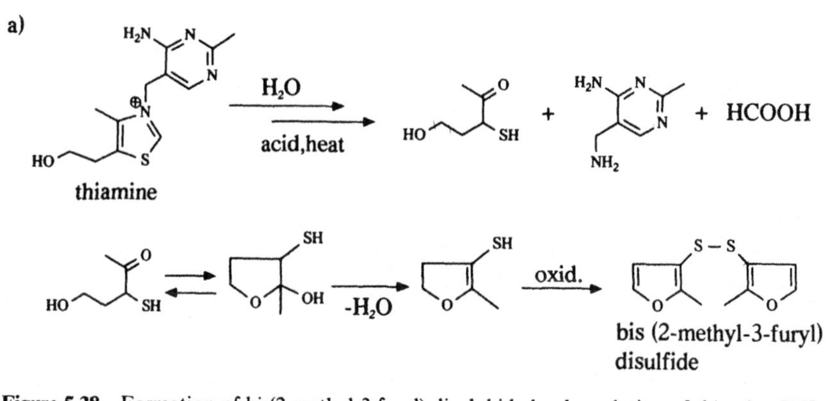

a)

thiamine

bis (2-methyl-3-furyl)
disulfide

Figure 5.38 Formation of bis(2-methyl-3-furyl)-disulphide by degradation of thiamine [78].

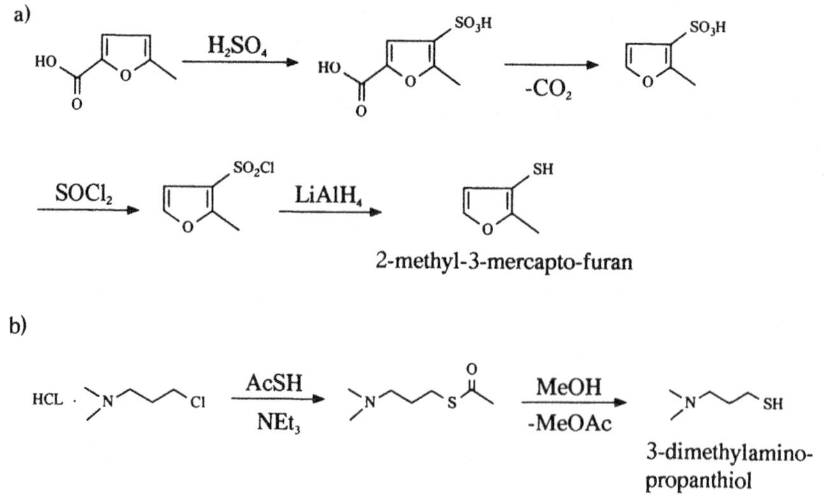

a)

2-methyl-3-mercapto-furan

b)

3-dimethylamino-
propanthiol

Figure 5.39 Synthesis of (a) 2-methyl-3-mercapto-furan [84] and (b) 3-dimethylamino-propanthiol [85].

the resulting flavour effects. Non-volatile derivatives of nucleotides and peptides as well as minerals are presumably reponsible for typical meat mouthfeel. Investigations using model systems have shown that a multitude of components are formed, but only a small range of them are really flavour significant in low concentration [81]. A perfect reconstitution of meat flavour with all its nuances seems to be practically impossible. That is why processed meat flavour 'bases' or extracts are preferred as building blocks for composition.

Bis-(2-methyl-3-furyl)-disulphide is described as having an extremely low threshold value [82]. A possible formation can be explained by thiamine decomposition (Figure 5.38) [78]. Investigations on the mechanism of thiamine degradation and analysis of the reaction products is still an active research area [83]. Figure 5.39 shows the chemical synthesis of 2-methyl-3-mercapto furan [84] and 3-dimethylamino propanethiole [85], representing quite different meat flavour perceptions.

5.2.3.12 *The fatty rancid flavour note.* The most potent examples of fatty rancid flavour notes are butyric and isobutyric acids, which are described as unpleasant, sour and repulsive. Caproic acid is weaker and rather cheesy rancid. Medium chain methyl branched fatty acids like 4-methyl octanoic and nonanoic acids are responsible for the off-flavour of mutton fat.

Long chain fatty acids are perceived as oily, waxy and soapy. The aldehydes cited in Figure 5.40 with their individual specific notes contribute to the fatty rancid perception depending on their concentration in different applications.

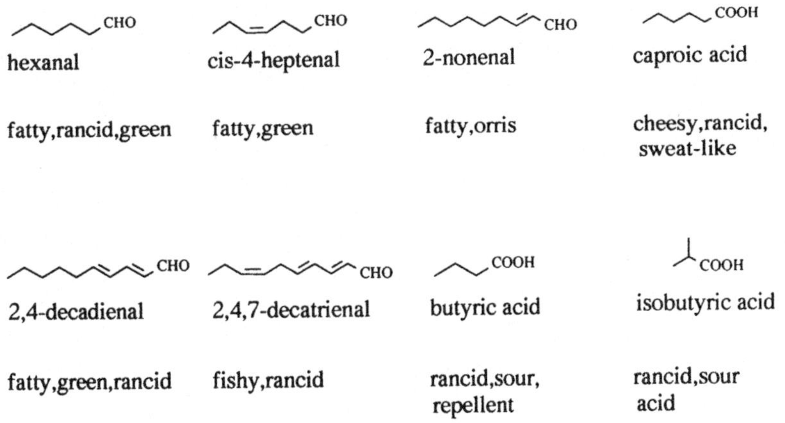

hexanal   cis-4-heptenal   2-nonenal   caproic acid

fatty,rancid,green   fatty,green   fatty,orris   cheesy,rancid, sweat-like

2,4-decadienal   2,4,7-decatrienal   butyric acid   isobutyric acid

fatty,green,rancid   fishy,rancid   rancid,sour, repellent   rancid,sour acid

Figure 5.40 Examples of fatty rancid flavour notes.

linoleic acid $\xrightarrow{\text{oxidation}}$ hexanal , 2-heptenal , 2,4-decadienal (E,Z and E,E) caproic acid , caprylic acid 9-oxo-nonanoic acid , 10-oxo-decenoic acid

linolenic acid $\xrightarrow{\text{oxidation}}$ 2-butenal , 2-hexenal , 2,4-heptadienal (E,Z and E,E), 2,4,7-decatrienal (E,Z,Z ; E,E,Z ; E,E,E) , caprylic acid , 12-oxo-10-dodecenoic acid 13-oxo-9,11-tridecadienoic acid ,

Figure 5.41 Formation of acids and aldehydes by autoxidation [42].

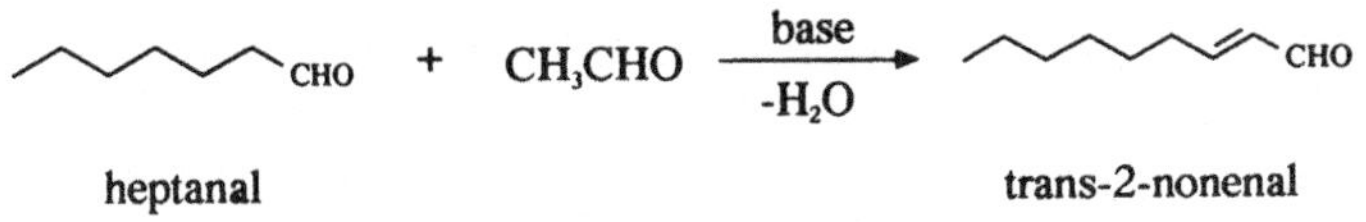

heptanal $+$ CH_3CHO $\xrightarrow[-H_2O]{\text{base}}$ trans-2-nonenal

Figure 5.42 Synthesis of trans-2-nonenal.

The same concentration of cis-4-heptenol can be differently evaluated in fat and fruit products.

Figure 5.41 shows the formation of some aldehydes and fatty acids generated by degradative fat oxidation of linoleic and linolenic acid, respectively [42]. The same products can be formed enzymatically, e.g. in fruits or vegetables, where they have desired flavour effects. In Figure 5.42 a standard synthesis of an α,β-unsaturated aldehyde trans-2-nonenal is outlined.

5.2.3.13 *The dairy buttery flavour note.* The dairy buttery flavour note varies from typical buttery notes (diacetyl, acetoin, pentandione) to sweet creamy fermented notes (acetol-acetate, δ-decalactone, γ-octalactone). Figure 5.43 shows some components representing this flavour type. It is also known that several aliphatic aldehydes and acids generated by lipid oxidation reactions contribute to the full flavour sensation. Polyoxygenated long chain fatty acid precursors as well as protein degradation products are presumed to be responsible for the fatty mouthfeel that occurs when butter is processed.

Figure 5.44 shows the biogenesis of acetoin and diacetyl, resulting from the well-known fatty acid metabolism. The synthesis of δ-lactones has already been mentioned (Section 5.2.3.2). γ-Lactones can be routinely synthesised according to Figure 5.45 [86].

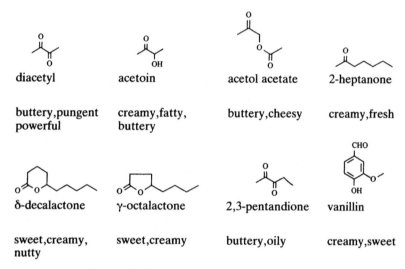

diacetyl	acetoin	acetol acetate	2-heptanone
buttery,pungent powerful	creamy,fatty, buttery	buttery,cheesy	creamy,fresh

δ-decalactone	γ-octalactone	2,3-pentandione	vanillin
sweet,creamy, nutty	sweet,creamy	buttery,oily	creamy,sweet

Figure 5.43   Examples of dairy buttery flavour notes.

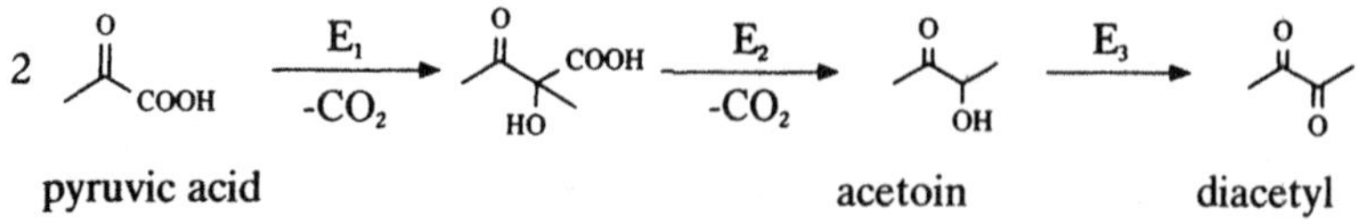

Figure 5.44   Biosynthesis of diacetyl and acetoin. E_1, E_2,..., different enzyme systems.

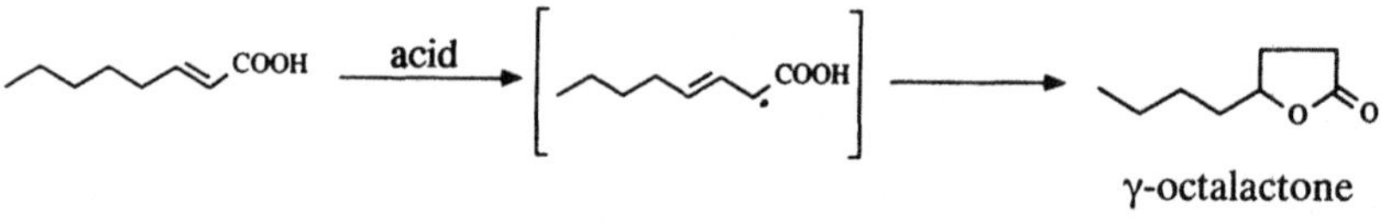

Figure 5.45   Synthesis of γ-octalactone [86].

5.2.3.14   *The mushroom earthy flavour note.*   The mushroom earthy flavour note is mainly represented by 1-octen-3-ol which is reminiscent of a typical mushroom flavour and by geosmin, representing the earthy part. A study of flavour constituents in mushrooms has shown that beside 1-octen-3-ol, a range of mainly C_8 components, saturated and unsaturated alcohols and carbonyls are responsible for typical mushroom flavour [87].

Other synthetic ingredients like 2-octen-4-one and 1-pentyl pyrrole are known to possess mushroom-like flavour notes (see Figure 5.46). 4-Terpinenol, 2-ethyl-3-methylthiopyrazine and resorcinol dimethyl ether with quite different chemical structures also have earthy flavour characters. Figure 5.47 shows the biogenetic pathway leading to 1-octen-3-ol, which has been only partially clarified [42]. The enzymatic oxidative degradation of

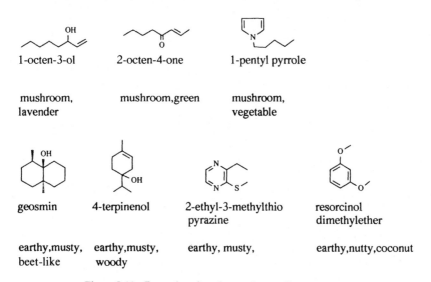

Figure 5.46   Examples of earthy mushroom flavour notes.

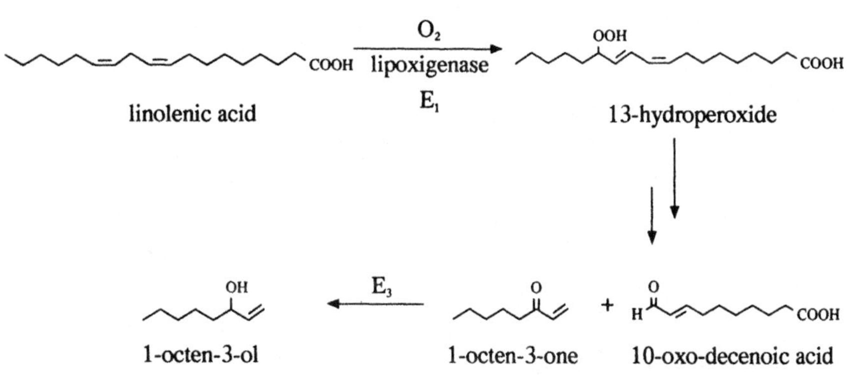

Figure 5.47   Biogenesis of 1-octen-3-ol [87]. $E_1, E_2, ...,$ different enzyme systems.

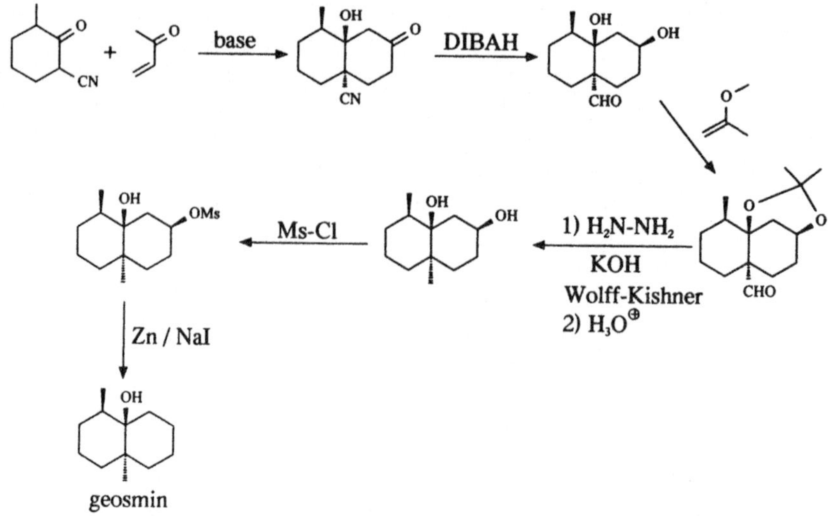

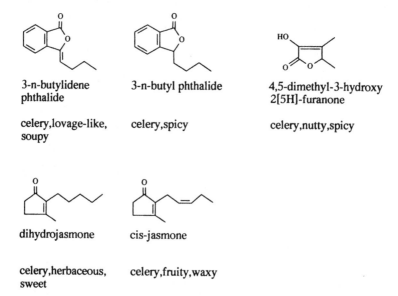

Figure 5.48 Synthesis of geosmin [88]. Ms, $CH_3SO_2^-$ (mesyl); DIBAH, $(i\text{-}Bu)_2AlH$.

linolic acid into C_8 and C_{10} components is observed only in mushrooms. Figure 5.48 shows a recent stereospecific synthesis of geosmin [88].

5.2.3.15 The celery soupy flavour note. The celery soupy flavour note is described as the warm spicy rooty odour, reminiscent of concentrated soup. Typical components are shown in Figure 5.49 (butylidene phthalide, butyl

3-n-butylidene
phthalide

celery,lovage-like,
soupy

3-n-butyl phthalide

celery,spicy

4,5-dimethyl-3-hydroxy
2[5H]-furanone

celery,nutty,spicy

dihydrojasmone

celery,herbaceous,
sweet

cis-jasmone

celery,fruity,waxy

Figure 5.49 Examples of celery soupy flavour notes.

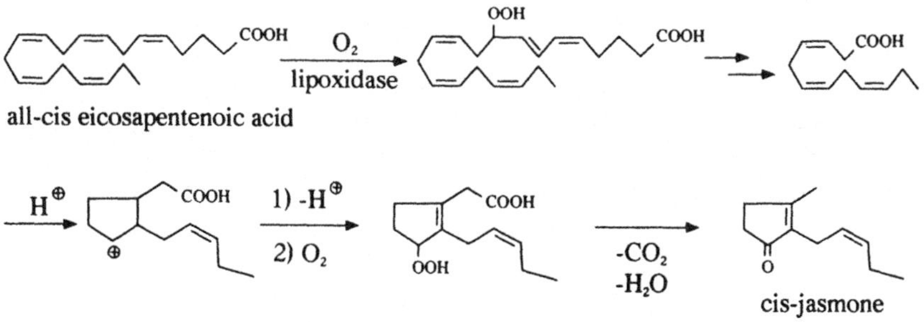

Figure 5.50 Biogenesis of *cis*-jasmone [89].

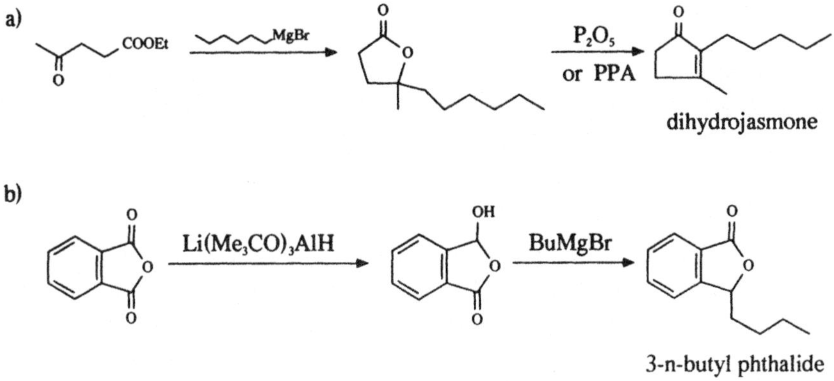

Figure 5.51 Synthesis of (a) dihydrojasmone [90] and (b) butyl phthalide [91]. PPA, Polyphosphoric acid.

phthalide, 4,5-dimethyl-3-hydroxy-2[5*H*]-furanone, dihydrojasmone and *cis*-jasmone). These notes differ from those described in Section 5.2.3.10. Both have some persistent enhancer character. The distinction to be made is that the celery notes are associated with vegetable applications, while the bouillon notes are appropriate in meat flavours.

4,5-Dimethyl-3-hydroxy-2[5*H*]-furanone is in its enhancing character comparable to 5-ethyl-4-methyl-3-hydroxy-2[5*H*]-furanone (Figure 5.49). *cis*-Jasmone and dihydrojasmone represent more the fruity sweet part of celery flavour. Figure 5.50 shows one of the first proposals for biogenetic formation of *cis*-jasmone [89]. Recent investigations led to the idea that a suitable polyunsaturated fatty acid is first cyclised via an allene oxide as an activated intermediate and secondly shortened to the proper length by β-oxidation [89a]. Figure 5.51 shows synthetic routes to dihydrojasmone and 3-*n*-butyl phthalide according to [90] and [91].

5.2.3.16 *The sulphurous alliaceous flavour note.* The sulphurous alliaceous flavour notes are vegetable specific notes, which are generally easy to recognise. The odour effect ranges from the simple unpleasant mercaptans (methyl mercaptan) through unsaturated short-chain garlic and onion compounds (allyl mercaptan, diallyl disulphide) to pleasant distinctly nuanced heterocyclic compounds. Examples therefore are 1,2,4-trithiolane for asparagus, lenthionine for mushroom (shiitake) and 2-methyl-4-propyl-1,3-oxathianes for passion fruit (Figure 5.52).

Sulphur containing components generally represent a very important class of flavourings with extremely low threshold values. They need very special handling because they often show instability (e.g. oxidation) and interactive

CH₃SH

methylmercaptane   dimethyl sulfide   dimethyl disulfide   allyl mercaptane

cabbage,diffusive   cabbage,unpleasant   vegetable,sulfury   onion,garlic,potent

diallyl disulfide   allylisothiocyanate   asparagusic acid   1,2,4-trithiolane

garlic,pungent   pungent,mustard,   crude asparagus   sulfury,green,
                       horseradish   chives

Figure 5.52 Examples of sulphurous alliaceous flavour notes.

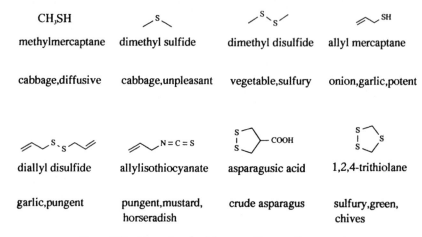

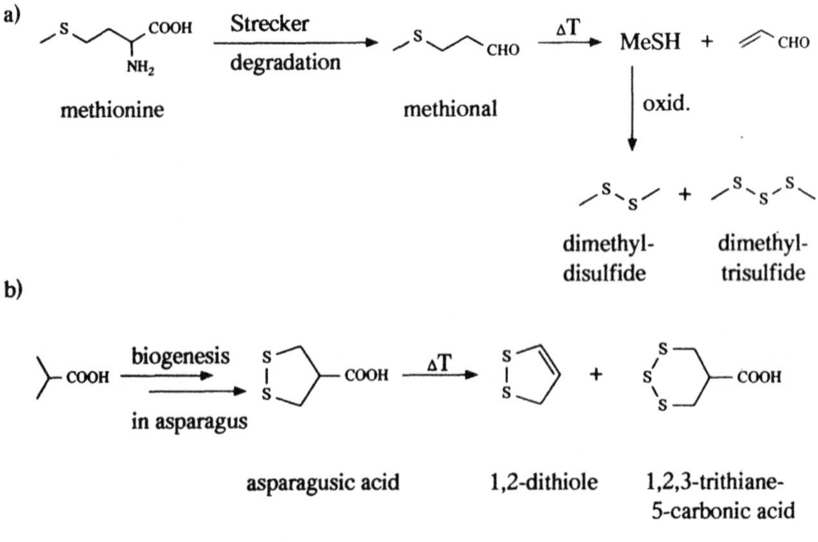

Figure 5.53 Formation of (a) aliphatic and (b) cyclic sulphur compounds [92].

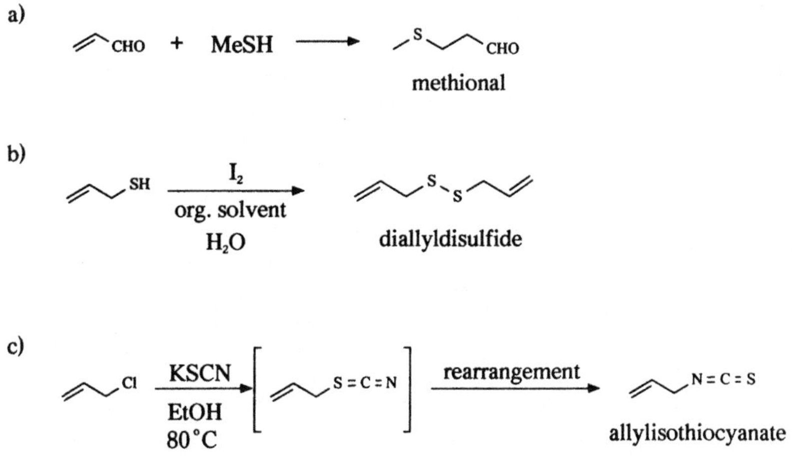

a)

$\diagup$CHO   +   MeSH $\longrightarrow$ $\diagup$S$\diagdown$CHO

methional

b)

$\diagup\diagdown$SH $\xrightarrow[\substack{\text{org. solvent} \\ H_2O}]{I_2}$ $\diagup\diagdown$S$\diagdown$S$\diagup\diagup$

diallyldisulfide

c)

$\diagup\diagdown$Cl $\xrightarrow[\substack{\text{EtOH} \\ 80°C}]{KSCN}$ $\left[\diagup\diagdown S=C=N \right]$ $\xrightarrow{\text{rearrangement}}$ $\diagup\diagdown N=C=S$

allylisothiocyanate

Figure 5.54 Synthesis of methional, diallyldisulphide and allylisothiocyanate.

reactions with other compounds in the flavour mixture. Sufficient knowledge about their behaviour is crucial for their correct application. The use of corresponding precursors is, therefore, often preferred. Sulphur containing flavour components are mostly formed during heat processing of plants and food materials. Figure 5.53 shows the formation of aliphatic and cyclic sulphur components. Strecker degradation of methionine leads to methional and by heat treatment to methyl mercaptan and related derivatives (Figure 5.53a).

The biogenetic precursor of asparagusic acid has been proved to be isobutyric acid [92]. 1,2-Dithiole and 1,2,3-trithiane-5-carboxylic acid, which contribute to the asparagus flavour, are presumed to be formed by thermal degradation (Figure 5.53b). The synthesis of methional, diallyldisulphide and allylisothiocyanate are outlined in Figure 5.54 as typical examples of organic sulphur chemistry [93].

5.3 Synthetic Flavour Ingredients and the Future

Synthetic flavour ingredients will certainly remain indispensable and will continue to be used, for many reasons, more in the future than the past. Figure 5.55 shows the increase in number of single flavour components identified in foods and beverages in the last 40 years. The largest increase is observed between 1965 and 1985 due to the progress in performance of new analytical techniques like GC and GC-MS-DS, HPLC, etc.

The use of these compounds (curve A) as potential flavouring agents depends mainly on the knowledge of their sensory properties, their commercial availability and toxicological behaviour. This is one of the reasons why only a small number of them are expected to be used in the flavour industry

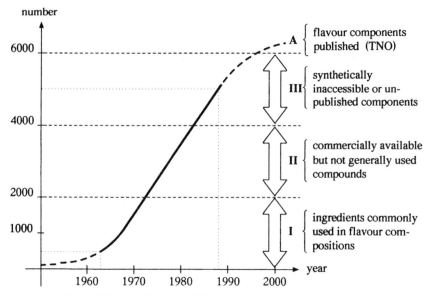

Figure 5.55 Development of flavour ingredients identified in food.

(section I). Although their number is growing constantly, they will never reach the potential maximum (section II) which represents approximately two thirds of the whole range of flavour substances actually known. A better understanding of their behaviour and interactions (single or in mixtures, including odour and taste threshold, stability, synergism, etc.) appears crucial and essential for the future.

At the same time a precise descriptive language using common food flavour specific terms (considering not only odours but also taste descriptions and threshold values) to describe flavour profiles, is needed world-wide to facilitate the dialogue between those involved in creation, application and marketing of flavourings and their customers.

A large number of interesting flavour substances, however, will remain synthetically and economically inaccessible or forbidden for safety reasons. Some of them can be considered as unpublished, company specific know-how (section III).

Flavour science and research will concentrate on special topics like non-volatile flavour precursors (glycosinolates) biotechnologically and Maillard generated flavour compounds (or 'bases') which can be used as single ingredient or as building blocks to replace or improve inaccessible flavour notes.

Flavour creation is a dynamic process which constantly evolves with progress in flavour chemistry. Considerable efforts have to be made today to detect and identify one single character impact substance occurring in the sub-ppb range. Sulphur containing flavour compounds with very low odour thresholds remain the most interesting field for future flavour compositions.

To use them optimally we have to understand not only their biogenesis but also their physiological effects on perception and organoleptic sensations. In this field, much work remains to be done since we are at present just scratching the surface. Very little is known about relationships between flavour chemicals and how stimulation occurs. Different model systems and molecular mechanisms [94] are proposed but not yet confirmed. Selected synthetic flavour ingredients will one day be able to play the role they are designed for, being used in quantity and quality, in the same way as if they were formed naturally.

The age of flavour design or flavour engineering postulated in the early 1960s after the appearance of gas chromatography can only begin when all the preliminary work is complete (some well-known companies are already well on the way to achieving this end). Computer aided selection of synthetic ingredients can help the flavourist in his composition or formulations to save time and money avoiding 'hit and miss' trials which are often caused by overhurried marketing demands to satisfy the needs of new short-life products observed on today's and future markets.

Even if the current trend of preference for natural ingredients continues, synthetically compounded flavours will continuously improve with regard to quality, commercial and ecological aspects. More investigations on minor compounds have to be carried out to avoid using global building blocks with all the risk they may include. A clarified and simplified legislation concerning synthetic flavour ingredients will be helpful in the future. First steps in this direction have already been initiated in Europe with a new EEC flavour directive [95], but corresponding basic regulations are requested world-wide to harmonise the needs of all consumers and the flavour and food industries.

6 Beverage flavourings and their applications

A.C. MATTHEWS

6.1 Introduction

Our requirement for liquid refreshment is as longstanding as the origins of the species homo sapiens. An examination of Figure 6.1 shows that at relatively low levels of fluid loss, e.g. 3%, impaired performance results. A 20–30% reduction in capacity for hard muscular work occurs with a moisture/fluid loss of 4%, heat exhaustion at 5%, hallucinations at 7% and circulatory collapse and/or heat stroke at 10% fluid loss. The required fluid intake for the average person in the arid areas surrounding the Red Sea is a staggering 8 litres per day.

With the very existence of life depending on our fluid intake, why do we not simply take water alone? The first reason is that water alone is not readily taken into human body systems, as it requires a certain level of carbohydrate and salt for rapid transfer across the brush border of the gut. The second reason is that water alone is uninteresting and unquenching in taste.

The importance of fluid intake at the correct osmotic pressure (a measure of tonicity and hence compatability with body fluids, the desired goal being iso-tonicity) is at the very origin of human nutrition. What was the reason for the original manufacture of local wines at alcohol contents of approximately 5% in the tropical, subtropical and arid areas of the world? It was simply to use the preservation properties of alcohol as a means of keeping grape juice from degrading to a non-potable state. In the process, rehydration fluid (water) was held in a microbiologically stable and acceptable state suitable for use over extended periods of time. This was the origin of the category of food products that has the universally accepted nomenclature, beverages.

In the early days of wine production, it was found that by using different grape juices and fruit juices in varying proportions a multitude of completely different tastes was achievable. The occasional use of only small quantities of certain fruits, vegetables, or herbs with strong characters (e.g. mango or thyme) in the blend of juices to be fermented would result in a completely different taste of the finished wine. This was probably the first regular use of the application of food flavourings to beverages.

Any substance which, when added to a beverage in quantities lower than 50% of the total volume, causes the flavour of the whole product to assume its

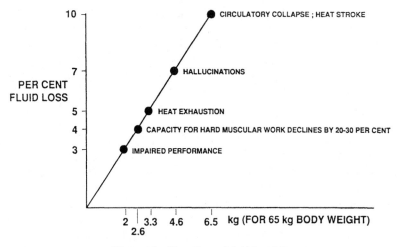

Figure 6.1   The effects of fluid loss [1].

own character may technically be called a flavouring. However, for the purposes of this chapter, we will consider such flavourings only in passing and concentrate our discussions on intense, relatively low volume use flavours.

One of the problems when trying to discuss the composition and application of food flavourings to beverages is the number and diversity of beverage industries around the world. An overview of some of the major sectors of these industries is presented in this chapter.

6.2   Categories of Beverages

From a technical point of view this diversity of beverage type and composition is considered so that flavour solubility and miscibility are not compromised when applying flavourings to beverages. For this purpose we will consider three broad categories of basic composition:

  (a) alcoholic beverages, subdividing into alcohol contents
     (i) lower than 20%
     (ii) higher than 20%
  (b) aqueous and aqueous/sugar beverages subdividing into sugar contents
       (i) lower than 2.5%
       (ii) higher than 2.5% and lower than 60%
       (iii) higher than 60%
  (c) mixtures of (a) (b), e.g.
       (i) liqueur
       (ii) low alcohol wine
       (iii) shandy or beer cooler

The composition of either naturally occurring or synthetically produced flavouring material must now be examined closely for compatibility with the category of beverage that requires flavouring. For example, some of the most effective and readily available natural flavouring materials are citrus essential oils released from the peel of citrus fruits during processing for juice extraction. These oils, as obtained from the fruit are almost insoluble in aqueous and aqueous sugar-containing beverages, are generally soluble in high alcohol containing beverages and have limited solubility (but greater miscibility) in aqueous beverages with sugar contents higher than 60%. These solubility/ miscibility characteristics are predictable from an examination of the composition of the oils, which shows a very high water insoluble terpene content. The composition of both lemon and orange oils is given together with information on essential extraction requirements [2] in Appendix I.

The best beverage composition for the natural whole citrus oil which has been neither solubilised nor processed is thus a high alcohol-containing beverage with a high sugar content. This is usually called a liqueur. Many famous names in this beverage category are renowned for their strong, fresh, authentic citrus characters that would be almost impossible to obtain via any other flavour route since the other methods which we shall consider for the processing of these and other such essential oils result in a diminution in authenticity of character unless this is compensated for in other ways.

6.3   Types of Flavourings for Beverages

Just as there are broad compositional categories for beverages so it is with flavourings for beverages. They result from either the form in which the concentrated natural or synthetic flavouring exists or result from the method or methods used to extract or manufacture the flavouring.

When the flavourist is compounding a flavour, he will start with an array of flavouring components of either natural or synthetic origin. These components will have different physical performance characteristics of dispersibility and solubility as well as individual flavour effects and interactive properties with other components. For natural flavouring materials these properties will, in part, be the result of the method of extraction and solvents used and will fall into three broad categories or types which then result in two basic performance properties:

(a) oil soluble flavourings subdividing into type by solvent
   (i) alcoholic
   (ii) non-alcoholic
(b) water soluble flavourings
(c) mixtures of (a) and (b), for performance reasons and generally operating at the limits of technical acceptability of the beverage application

The synthetic components will also carry with them performance characteristics which may be categorised in a similar way.

6.4 Methods of Extraction, Solubilisation and Concentration of Flavourings

Since water (alone)-soluble flavourings are by far the smallest category for the
flavouring of beverages it would seem sensible to consider these before oil-
soluble and mixed-solvent flavourings. Their acceptability and ease of use in
the aqueous non-alcoholic beverage category and industry (as defined in this
chapter) would make them the preferred route for flavour application if it were
not for the limitation they impose on flavour range and intensity. The biggest
and most widespread extraction and use of water-soluble flavourings is instant
coffee.

6.4.1 *Extraction of Coffee Flavour and Manufacture of the Instant Product*

Instantly soluble coffee is prepared in a series of column extractors in which
the finely ground roast beans are extracted with counter currents of water
under pressure at 175°C (350°F) for a period of about 4 h. This results in an
almost complete recovery of soluble solids from beans and gives about 45–
50% of extract. The initial concentrate is then either spray dried to give the
familiar free-flowing powder or freeze dried and agglomerated to give a
product looking more like ground coffee. To improve the 'freshly perked or
filtered' coffee aroma, about 0.2% of recovered coffee oil is incorporated into
the product by either spraying onto the surface of the powder or by
incorporation into the concentrate prior to freeze drying. The resulting
products are packed under inert gas into glass jars to protect flavour freshness
and intensity during storage prior to use.

In this example of what appears on the surface to be a fully water extracted
and soluble flavouring system, we can already see that certain very important
and significant volatile (taste and, in particular, aroma) components are lost if
only the water extract alone is used. This is because when any flavour
extraction is carried out at temperatures of 100°C and above (at atmospheric
pressure) any essential oils and, dependent upon temperature, resinous
materials will be volatilised and carried over in the steam released. The
effectiveness of this volatilisation, which must be compensated for when
formulating the finished beverage, will be dependent upon the particle size of
the material under extraction since this has to be small enough to allow good
steam penetration. This can be seen even more clearly in the next most
commonly perceived 'water-soluble or extracted' group of flavourings,
concentrated fruit compounds and flavouring extracts where the flavour of the
fruit is divided unequally between the juice and the volatile and oil
components.

6.4.2 *Flavourings Extracted from Harvested Fruits*

Fruits are the matured ovary of the plant or tree and may be with or without seeds and sometimes with the flower still attached. The wall of the fruit developed from the wall of the ovary is called the pericarp and may be either dry or fleshy; it is this edible fleshy part which forms most of the varieties we call 'fruits'. Nuts are also fruits and they will be covered separately for their application to beverages and, in particular, the manufacture of cola nut extracts and flavourings. Botanically fruits may be classified as in Appendix II [3].

This is the type of classification and way of thinking about fruits that a beverage formulator will pursue as it follows the perception and requirements of the consumer. A flavourist, however, will be more interested in the value, diversity and intensity of the flavouring materials as library items from which to draw when compounding flavourings. He or she may be less interested in the replication of the whole fruit flavour in the finished beverage, this being the job of the applications technologist. The processing of fruit for the purpose of extracting juice is a complex area of technology involving the use of enzymes, preservatives, different expression techniques, de-activation of natural enzymes, heat processing and flavour retention, collection, and recovery techniques (Chapter 4). A very full account of fruit processing has been also been given by Tressler and Joslyn [4] and will only be covered in outline here as far as it relates to flavour extraction, formulation, beverage application and their interactions.

Very few flavourings which may be obtained from natural materials for beverages can be extracted and dissolved by water alone. In citrus fruits the juice cannot possibly be characteristic of the whole fruit since so much of the character is contained in the oil component in the peel.

In so-called berry fruits, the method of hot enzyming and extraction ensures that the volatile organic components will be lost during concentration in the steam/water phase unless steps are taken to recover them. The characteristic volatile aroma that is responsible for the flavour profile of most freshly pressed juices is present at levels typically around only 700 parts per million and considerably less in heat treated juices. In the extraction of flavouring materials from the juicing process, it is intended that no components of potential value are wasted and that the valuable non-aqueous phase materials are collected. These components are insoluble and form an unstable emulsion when applied to most liquid beverages.

6.4.3 *Extraction and Use of Oil Soluble Flavourings*

Oleoresins, tinctures and extracts have already been discussed in Chapter 3. In this section, discussion is restricted to the methods of their application to beverages and where necessary (e.g. cola nut extract), specific methods of extraction which result in products compatible with aqueous, and aqueous sugar-containing beverages, in addition to beverages containing less than 20% alcohol.

6.4.3.1 *Solvent extraction.* If normal distillation techniques are considered too harsh for the extraction of more delicate or sensitive flavourings such as floral notes or cola nuts, then solvent extraction may be employed. The collected flower or finely ground nuts are placed in tiers on perforated plates in a suitable extraction vessel and the selected solvent (usually about 60% alcohol or propylene glycol in the case of cola nuts) is allowed to percolate until the raw material is exhausted. The resulting solvent containing extract may then be used directly or reduced to a tincture or soft extract by distillation to recover solvent. This may take place at elevated temperature and an example is given in Table 6.1.

6.4.3.2 *Essential oil extraction by pressing.* This is the method used to extract essential oils from fruits with peels which have a naturally high oil content such as the citrus fruits of orange, lemon, lime, etc. It is the method used by most of the world's citrus juice processors using FMC-type extractors. Oil released from the peel during processing is washed away from the fruit with a spray of water. The oil and water are subsequently separated by centrifugation. Owing to the incompatibility of essential oils with aqueous phase systems (see above) the highest quality essential oils are those with minimal (or the least) contact with water.

Oils may also be extracted by steam distillation but will still require the use of solvent extraction and/or subsequent use of solvents to make the resulting flavouring practically usable in aqueous sugar beverages and beverages containing < 20% alcohol.

Ethyl alcohol or isopropyl alcohol (although imparting different flavour characters in their own right) are widely used as solvents for beverage flavourings. They are normally compatible either used singly or in combination with each other or water with aqueous, aqueous sugar-containing and alcohol (< 20%)-containing beverages. In parts of the world where the use of these or other alcohols is either undesirable or legislatively not permitted,

Table 6.1 Cola nut flavour

		Quantity (g)
1.	Finely divided (into small particles) cola nuts	400.00
2.	Add propylene glycol	800.00
3.	Heat and maintain at 30° to 35°C for at least 8 h	
4.	Gently distil under vacuum to reduce distillate to	100.00
5.	Remove vacuum and add sufficient water (approx. 900 g)	
	to give total essence of	1000.00

Table 6.2   Lemon essence

1.	Cold pressed lemon oil	150.00
2.	Cold pressed lime oil	50.00
3.	Alcohol, 95%	550.00
4.	Water	250.00
	Total	1000.00
5.	Agitate and stand 24 h to separate terpenes	− 150.00
	Yield	850.00
6.	Add alcohol, 95%	150.00
	Lemon essence, terpene-free (filter if necessary)	1000.00

[a]May be partially replaced by, or supplemented with lemon grass oil which is the main source of natural citral (approx. 80% geranial and 20% neral).

alcohol as an extractive solvent must be removed by distillation and the resultant extract redispersed in other solvents/diluents such as propylene glycol or tri-acetin (glyceryl triacetate).

A traditional method for the preparation of flavourings from essential oils is that of 'washing the oil' or oils. This involves mixing the essential oil/s with alcohol, water and sometimes other solvents. The mixture is allowed to stand for up to 24 h for separation to occur with the subsequent removal of insoluble terpenes. The resulting essence is then further solubilised with alcohol. A typical formula is given in Table 6.2.

6.5   Beverages Based on Ginger

An example of a flavour which has application across all three of the previously defined beverage categories (including the subdivisions) and the

Table 6.3   Examples of beverages based on ginger

Beverage[a]	Aqueous/sugar (% sugar content)	Alcoholic (% alcohol content)
American ginger ale	10–12	–
Diet American ginger ale	< 2.5	–
Original ginger ale	4–8	–
Ginger beer flavour*	6–13	–
Ginger beer*	6–13	1
Ginger cordial	23–60	–
Ginger cordial	23–60	10
Ginger wine	30–40	10–20
Ginger liqueur	60	40

[a]All products listed are normally of clear appearance. However, those marked with * are often cloudy since they use an emulsion containing oil (or other optically reflective droplets) finely divided and stabilised in the aqueous phase.

three flavour categories (including the two broad performance types) is ginger (Table 6.3).

Every flavourist or formulator who has worked with ginger or its derivatives knows how dependent the final product is on the origin of the natural raw material used. The flavour ranges from lemony, spicy, through aromatic, earthy, harsh, to camphoraceous and the methods of flavour extraction employed (e.g. to give sweet, soft ginger flavour through to fiery, hot and nose tingling). These different flavour effects may be achieved by the use of different proportions and combinations of raw material and also by the selection of both aqueous and non-aqueous phase extracts and by the skillful manipulation of the product formulation into which these materials are applied.

6.5.1 Manufacture of Ginger Extract

A formula and process for the manufacture of a general purpose ginger extract is given in Table 6.4 [3]. The flavour of the extract so achieved may be modified in a number of ways in order to achieve the required performance target. For a flavourist this will be to obtain the maximum number of flavouring components to use in a variety of different product applications including beverages. By use of different extraction methods to obtain very different flavour characteristics, the general extract can be widely modified, e.g. by use of oleoresin of ginger (see Chapter 3).

The flavourist's perception of the required performance criteria of the flavouring will usually be physical (appearance and compatibility) and organoleptic stability in the general flavour target area. This is a result of the

Table 6.4 Basic ginger extract

Extraction of:
  (1) 132 kg Jamaica ginger, comminuted, with:
      480 kg or 580 litres heated water of about 90°C, circulate the liquid over the ginger for 10 min, then let the mixture stand for 24 h to cool

  (2) To the mash of (1) is then added:
      285 kg or 420 kg alcohol, 95%; the menstruum is to be circulated for 10 min to obtain uniform alcohol content of 40% strength

  (3) After 3 days extraction during which the menstruum has been circulated twice daily, the liquid is drained off and yields approx.

  (4) 670 kg ginger extract; it is used in (5)
      226 kg remaining mash is mixed with
      150 kg water, and distilled at atmospheric pressure, to yield
      110 kg flavour distillate of about 25% alcohol content, it is used in (5).

  (5) Mix:
      670 kg ginger extract of (3) with
      110 kg flavour distillate of (4), to yield

Total
  1720 kg or 1000 litres basic ginger extract

---

single most difficult problem that has to be overcome with beverages, i.e. the compatibility of most flavourings with their varying liquid compositions.

An examination of how the basic ginger extract may be modified to achieve different desired flavour effects for different beverage applications will serve to demonstrate how the flavourist and applications technologist can contribute to the development of different flavourings. The two products considered are:

(a) 'Original' (hot) dry ginger ale (carbonated beverage) and
(b) 'American', 'Canada', or 'pale' (soft and sweet) ginger ale (carbonated beverage).

6.5.2 'Original' (hot) Ginger Ale

The formula required to prepare flavouring of the correct character for 'original' (hot) dry ginger ale is shown in Table 6.5. It can be seen that the flavourist is achieving the 'drier' and 'hotter' ginger taste by the use of oleoresins of both capsicum and ginger and is 'deepening' or 'broadening' the resulting flavour with the use of ginger, orange, lime, mace and coriander oils. The use of rose oil in this context must be very carefully controlled since an excess will render the flavour 'out of balance'. This expression, 'out of balance', is used a great deal in the flavour and flavour application industries and we will return to it in a different context later.

The product formulation given in Table 6.6 is typical of the type of carbonated beverage of the 'original' or 'dry' ginger ale category. Here it can be

Table 6.5 Original ginger ale essence

Mixture of:

9.4	ml oleoresin ginger[a]
0.95	ml oleoresin capsicum
1.9	ml oil of ginger[a]
1.9	ml oil of orange, cold pressed
1.9	ml oil of lime, distilled
0.25	g oil of mace
0.25	g oil of coriander
200	ml alcohol, 95%
400	ml propylene glycol
12.5	g magnesium carbonate
	water

Procedure:

(1) Mix oleoresins with 150 ml alcohol (95%) and 220 ml propylene glycol; add 6.85 g magnesium carbonate, mix 1 h, and then let stand overnight; decant and filter

(2) Mix the oils with 66.15 ml alcohol (95%) and then with the filtered mixture of (1); afterwards add to it 20 ml propylene glycol and 20 ml water to make 1000 ml essence

(3) A few drops of oil of rose are sometimes added to the essential oil mixture to give it a distinctive character

[a] May partially be supplemented or enhanced by the use of a proportion of 'basic ginger extract' shown in Table 6.6.

Table 6.6 'Original' or 'dry' ginger ale (carbonated beverage): materials and quantities for 1000 litres finished product

Material	Quantity	Units
Granulated sugar	38.9	kg
Sodium saccharin dihydrate[a]	0.06	kg
Sodium benzoate	0.174	kg
'Original' ginger ale essence and adjust to taste	0.400	litres
American ginger ale or tonic water essence	~0.160	litres
Citric acid anhydrous (to give 0.25% w/v CAMH)	2.30	kg
Caramel MW3	0.21	litres
Water to 1000 litres	~970	litres
Carbon Dioxide to 4.5 vols. bunsen (v/v)	1000	litres

Directions/special notes
Make into a suitable syrup, dissolving the solid ingredients in water before addition; dilute the caramel 1:1 with water before addition; after adequate mixing use the syrup for bottling by diluting one volume with an appropriate volume of water, adding carbon dioxide in a suitable manner

Analytical characteristics

Refractometric Brix (% w/w at 20°C)	4.0 ± 0.3
will increase if saccharin replaced by sugar to	7.0 ± 0.3
Acidity as citric acid monohydrate (% w/v)	0.25 ± 0.03

Calculated sweetness

(1) Due to sugar (% w/v)	3.90
Due to saccharin (% w/v)	3.00
Total	6.90
or, any combination up to	
(2) Due to sugar (% w/v)	7.00

seen that the applications technologist is achieving the 'dryness' of taste by keeping the sweetness low (total sweetness of 6.9% w/v compared with 10–12% w/v for most products of this type) and the acidity at an average level (0.25% w/v as citric acid monohydrate) and hence is obtaining a low sweetness/acid ratio which will make the product taste 'sharp' and relatively low in sweetness. The use of sodium saccharin in this context (where permitted legislatively and by consumer acceptance) is to give the product an even 'harsher' taste since a lower sugar content used in combination with this sweetener will remove some of the perceived 'body' of the beverage. The inherent 'thinness' or 'strident' nature of the sweetness of saccharin will further enhance the 'dryness' of taste and allow the strong ginger and heat characteristics of the flavouring components to become apparent. The total effect is to intensify the 'dry', and 'hot' ginger character of the beverage. The use of sweetness/acid balance achieves a better 'original' ginger character than by the use of flavouring alone at higher addition levels combined with higher sweetness. If this product formulation has intensified the 'hot' and 'dry' characters of the product too much for consumer acceptance (this information should always be gathered by correctly structured consumer taste testing/research), slight reduction of the level of the 'original' ginger ale essence and inclusion of a small proportion of either the 'softer', more 'citrus' 'American' or 'pale' ginger ale essence or a citrus blend such as that used in Indian tonic water, can be used beneficially.

The flavourist may in this way achieve through flavouring composition the same effect as the applications technologist does by varying the composition of the beverage.

6.5.3 'American' or 'Pale' Ginger Ale

A formula for the manufacture of a flavouring suitable for 'American' or 'Pale' ginger ale is given in Table 6.7. An examination of the formula for the manufacture of 'American' or 'Pale' ginger ale essence shows obvious differences from that for the 'original' (hot) ginger ale essence. Oil of rose is used as a small but required part of the character of the flavouring since in this type of product floral notes are necessary to obtain the 'balance' of required taste. It can also be seen that a 'softer', 'sweeter', less 'hot' product is achieved by omission of oleoresin capsicum and the use of much greater quantities of citrus oils, including bergamot oil with its much 'sweeter', 'softer', and 'fragrant' character.

The product formulation given in Table 6.8 is typical of the type of

Table 6.7 'American' or pale ginger ale essence

Mixture of:

0.5 g	oil of rose
0.5 g	phenylethyl alcohol
9.5 g	methyl nonyl acetaldehyde 50%
22.0 g	oleoresin of ginger
22.5 g	oil of ginger
27.0 g	oil of bergamot
246.0 g	oil of orange, cold pressed
300.0 g	oil of lemon, cold pressed
372.0 g	oil of lime, distilled
5000.0 g	alcohol, 95% then add
7000.0 g	distilled water and mix.

Procedure:
(1) The mixture is then poured into a separator, and left covered in a cool place for 48 h
(2) After the terpenes have settled on the surface of the mixture, the clear extract is taken off
(3) The remaining terpenes are mixed well with
    1500.0 g alcohol (95%) and to it added
    3500.0 g distilled water
    mix well for 5 min then pour the mixture into a separator; after 48 h take off clear extract and discard the terpenes
(4) Extract mixtures of (2) and (3) are united, and agitated for 5 min before the compound is poured into a separator; there the mixture is left for 48 h; the terpenes will separate to the surface; the clear extract is taken off, and the terpenes are discarded

Table 6.8 'Original' or 'pale' ginger ale (carbonated beverage): materials and quantities for 1000 litres finished product

Material	Quantity	Units
Granulated sugar[a]	90.0	kg
Sodium benzoate (give 145 mg/kg benzoic acid)	0.177	kg
Sodium citrate	0.21	kg
Lemon essence	0.10	litres
American ginger ale essence	3.25	litres
Caramel MW3	0.080	kg
Citric acid anhydrous (to give 0.15% w/v CAMH)	1.37	kg
Water to 1000 litres	~ 940	litres
Carbon dioxide to 4.5 vols. bunsen (v/v)	1000	litres

Directions/special notes

Make into a suitable syrup, dissolving the solid ingredients in water before addition; dilute the caramel 1:1 with water before addition; after adequate mixing, use the syrup for bottling by diluting one volume with an appropriate volume of water, adding carbon dioxide in a suitable manner

Analytical characteristics

Refractometric Brix (% w/w at 20°C)	9.0 ± 0.3
Acidity as citric acid monohydrate (% w/v)	0.15 ± 0.03

Calculated sweetness

(1) Due to sugar (% w/v)	9.00
or, any combination up to	
(2) Due to sugar (% w/v)	5.00
Due to aspartame (% w/v 200)	4.00
Total	9.00

[a] May be partially replaced by Nutra Sweet up to 0.20 kg.

carbonated beverage of the 'American' or 'pale' ginger ale category. Here it can be seen that the applications technologist achieves a 'sweeter', 'softer' character by keeping the usual level of sweetness for a carbonated beverage (total sweetness of 9–10% w/v compared with 7% w/v for the 'original' or 'dry' ginger ale) in combination with a relatively low level of acidity (0.15% w/v C.A.M.H.) and hence obtaining a high sweetness/acid ratio which will make the product taste 'soft', 'full-bodied' and relatively high in sweetness. The use of either all sugar or a high quality intense sweetener like aspartame helps the good perception of 'body' and 'roundness'.

In this example, the applications technologist supplements the efforts of the flavourist in achieving the required 'softer', 'sweeter' taste with the choice of a different sweetness/acid ratio and also in obtaining the required citrus blend with the use of additional lemon flavouring. Addition of extra floral notes and the use of other citrus components such as orange, lime or bergamot are also at the discretion of the product formulator.

A specialist area of flavouring/product formulation interaction is that of 'diet', 'low calorie' or 'sugar-free' carbonated beverages. The intense sweetener aspartame has increasingly gained acceptance for use in this area and is renowned for its interaction with flavourings and dependence upon shelf-life management and correct product formulation and composition.

An example of a 'sugar-free' 'American' type ginger ale product formulation is given in Table 6.9. The flavourings which the product formulator selects for use in such a product must be carefully evaluated and assessed for interaction with the sweetener system. In this example it is almost impossible for a flavourist to compound the correct flavouring without the specific help and interaction of the application technologist.

An examination of the sugar-free 'American' style ginger ale product formulation shows how the application technologist has compensated for loss of body and sweetness which results from the removal of 10% sugar by the use of a combination of polydextrose (provides body but no sweetness) and aspartame (provides sweetness but no body) with buffered low acid content to maintain a high perceived sweetness/acid ratio. This combination does, however, have a 'dampening' effect on the taste from the flavourings used in the sugar-containing analogue of this product. It is necessary to compensate for this effect by the use of extra levels of these same flavours and by the incorporation of a low level of 'original' (hot) ginger ale essence to restore the 'bite' of the product taste.

Table 6.9 'Sugar-free' American style ginger ale (carbonated beverage): materials and quantities for 1000 litres finished product

Material	Quantity	Units
Aspartame (NutraSweet)	0.50	kg
Sodium benzoate (give 145 mg/kg benzoic acid)	0.15	kg
Sodium citrate	0.25	kg
Polydextrose (Pfizer)	50.00	kg
American ginger ale essence	3.50	litres
Lemon essence	0.15	litres
Original ginger ale	0.15	litres
Caramel MW3	0.080	kg
Citric acid anhydrous (to give 0.15% w/v CAMH)	1.37	kg
Water to 1000 litres	q.s.	litres
Carbon dioxide to 4.5 vols. bunsen (v/v)	1000	litres

Directions/special notes
Make into a suitable syrup, dissolving the solid ingredients in water before addition; dilute the caramel 1:1 with water before addition; after adequate mixing, use the syrup for bottling by diluting one volume with an appropriate volume of water, adding carbon dioxide in a suitable manner

Analytical characteristics
Refractometric Brix (% w/w at 20°C)   9.0 $\pm$ 0.3
Acidity as citric acid monohydrate (% w/v)   0.15 $\pm$ 0.03

Calculated sweetness
(1) Due to aspartame (% w/v $\times$ 200)   9.00

If product formulator and flavourist were to work together, it may be possible to formulate a flavouring which would encompass many of these requirements for use specifically in sugar-free beverages. It is the skill of the applications technologist, however, in combining different flavourings to achieve the same result that gives many products their uniqueness and makes the beverage producer independent of any one flavour manufacturer.

6.6   Formulation of Beverages

6.6.1   *General Principles*

The general approach to the formulation of beverages is applicable across the wide range of products which fall into this category. These general principles apply to aqueous sugar-containing, alcohol-containing and powdered, or dried, beverages as long as the effects the different processes employed in the manufacture of the different products are taken into account when formulating. As an example, the flavouring of milks or milk-based beverages where the high temperatures employed in the UHT process are necessary for the long-life microbiological stability of the product, completely destroy the effect of certain (e.g. strawberry) natural flavourings. Nature identical versions of the sensitive flavourings are similarly, although to a lesser extent, affected by this loss of flavour. Artificial flavourings will normally withstand this type of process without loss of flavour character, intensity, or acceptance.

A strawberry natural flavouring extract is shown in Table 6.10 and the formula of a milk-based product which might use such a flavouring is shown in Table 6.11. A heat stable artificial strawberry flavouring formulation is given in Table 6.12 for comparison.

There are only a limited number of things that the applications technologist can do to try and overcome this problem such as adjustment of pH, use of anti-oxidant and an investigation of the process to see if it would be possible to add this heat sensitive natural flavouring in an aseptic way after pasteurisation of the product. If the flavourist is working with the beverage formulator there is the possibility of producing a WONF (with other natural flavours) strawberry essence that could have improved stability to heat. Heat instability of a natural flavouring could, for example, be the result of the initial process employed for the extraction of the natural flavouring from the fruit. A strawberry WONF flavouring of the type that may be used is given in Table 6.13 for information.

6.6.2   *Principal Components Used in the Formulation of Beverages*

When gathering together the flavourings to be used in the formulation of a beverage or preparing the brief for a flavourist to manufacture a speciality flavouring for that particular application, the product formulator will already have decided upon the general structure of the beverage formulation. This

Table 6.10 Strawberry fruit juice and flavouring extract. This procedure describes the production of a strawberry juice flavour from frozen strawberries

Procedure	Quantity	Units
A. Production of strawberry juice and flavouring extract		
(1) Defrosted whole Marshall strawberries	1000	kg
Granulated sugar[a]	200	kg
Pectinesterase (pectinol)	2.00	kg
Ethyl alcohol (95% v/v)	200	litres
This mix is comminuted using the Fitzpatrick machine with a no. 4 sieve and then pressed		
(2) The pressed pomace is mixed with water and distilled at atmospheric pressure	100	litres
Yield of distillate at approximately 60% v/v alcohol	14	litres
(3) Juice from (1) is mixed with distillate from (2) to give flavouring extract		
Approximate yield	450	litres
This extract has an alcohol content of 19–20% v/v; it is important that the level is more than 18% v/v alcohol; below this strength the juice will ferment rapidly; 1 litre expressed juice is obtained from approximately 1 kg of strawberries without sugar (0.85 kg with sugar)		
B. Strawberry fruit flavour		
Optimum flavour is obtained by vacuum distillation of the expressed juice from no more than 2 kg fruit per litre; use of more than 2 kg per litre yields an inferior product. This formula for strawberry fruit flavour is derived from the extract above, produced without added sugar		
(1) Strawberry juice extract ((3) above) expressed from approximately 1 kg fruit per litre with alcohol content 19% v/v	200	litres
This extract is concentrated under vacuum to produce a first fraction distillate at approximately 50% v/v alcohol		
Yield of first distillate	28	litres
(2) A second fraction produces distillate at approximately 30% v/v alcohol	72	litres
This is redistilled to higher alcohol content and used in the next production of flavour		
(3 A third, non-alcoholic fraction to be used in the next production batch in place of water, where it is mixed with pressed pomace ((2) above) prior to distillation		
Yield stage three distillate	64	litres
(4) Strawberry concentrate remaining	36	litres
(5) Formula for strawberry fruit flavour		
Cooled strawberry concentrate ((4) above)	36	litres
Add to the concentrate in the still to avoid loss of material		
Expressed strawberry juice and flavouring extract ((3) above)	36	litres
First fraction flavour distillate ((1) above)	28	litres
Yield of finished strawberry fruit flavour containing approximately 19% v/v alcohol	100	litres

[a] Optional ingredient.

Table 6.11   Strawberry milk-based beverage: materials and quantities for 1000 litres finished product

Material	Quantity	Units
Skimmed milk	400.00	litres
Strawberry purée	400.00	litres
Sugar	80.00	kg
Pectin	5.00	kg
Strawberry essence	0.40	litres
Citric acid	q.s.	to pH 4
Water to 1000 litres	q.s.	litres
Total	1000	litres

Analytical characteristics

Refractometric Brix (% w/w at 20°C)	12.5 ± 0.5
pH	4.0

Process details
(1) Mix the ingredients in the order listed taking care to disperse fully all of the small addition items; carefully add the flavouring and citric acid, measuring the pH during addition
(2) Ultra high temperature process this product by heating to 132°C for a minimum of 1 s prior to cooling and filling into appropriate containers

Table 6.12   Artificial strawberry flavour (wild strawberry)

Mixture of:

0.80 g	ethyl heptylate
0.80 g	oil of sweet birch
2.10 g	aldehyde C_{14}
2.40 g	cinnamyl isobutyrate
2.60 g	ethyl vanillin
3.00 g	Corps Praline (trade name) dissolved in
3.20 g	cinnamyl isovalerate
3.40 g	dipropyl ketone
5.00 g	methyl amyl ketone
6.00 g	diacetyl
21.20 g	ethyl valerate
23.15 g	aldehyde C_{16}
43.20 g	ethyl lactate
100.00 g	alcohol, 95%
783.15 g	propylene glycol
1000.00 g	Total

means that he or she will have already conducted an in-depth survey of the legislative framework into which the beverage will fit and will have included information on local 'custom and practice' which may be different from that allowed legislatively. It will therefore be assumed for the purpose of the discussion that this most vital of the beverage formulator's tasks is completed and the actual formulation only is considered.

The beverage formulator has a range of raw materials from which to draw in

Table 6.13   Strawberry WONF (yield 1000 g)

Ingredients:

500.00 g	alcohol, 95%
2.00 g	oil of rose
2.80 g	oil of jasmin
1.00 g	oil of cassie
1.00 g	oil of wintergreen
0.50 g	oil of lovage
2.50 g	oil of valerian
0.02 g	oil of celery
0.10 g	oil of coriander
520.00 g	distilled water

Procedure

(1) Agitate extract mixture of alcohol and oils, then add 520 g distilled water and continue agitating; transfer mixture into separator and leave it covered overnight; the terpenes will separate to the surface, while the extract below will be clear
(2) Take off the clear extract, discard the separated terpenes

Table 6.14   Principal components for the formulation of beverages

Ingredient	Contribution
Sugars	Flavour, sweetness, mouthfeel body, fruitiness, nutrition, facilitate water absorption (appearance and preservation in syrups)
Fruit/extract/milk/other characterising ingredient (e.g. glucose syrup, spring or mineral water, etc.)	Flavour, body, appearance, (nutrition)
Nutrient additions including salts	Nutrition: ascorbic acid and tocopherols are anti-oxidants also; controlled absorption of sugars and water
Acids	Flavour, antimicrobial effect
Flavours	Flavour, body, appearance;
Artificial sweeteners	carotene and riboflavin
Colourings	colourings are nutrients also
Emulsifiers and stabilisers	
Anti-oxidants	Improved flavour and vitamin stability
Preservatives	Antimicrobial effect; sulphite also has antibrowning and anti-oxidant effectiveness
Acidity regulators	Improved dental safety; reduced can corrosion; body
Alcohol (e.g. beers, wines, spirits, etc.)	Body, mass, solvent, carrier, flavour, mouthfeel, bite, punch, flavour potentiator or releaser
Water	Bulk and mass; solvent carrier; thirst quenching

Table 6.15 Comparison of sweeteners used in beverages [8]

Carbohydrates (100% solids base)	Taste characteristics					
	Sweetness intensity (10% sucrose)	Sweetness quality	Time profile	Associated taste	Mouthfeel, body	Enhancement of fruitiness
Sucrose	1.0	Full rounded	Fast, slight linger	None	More than other carbohydrate	Good
Invert sugar		Close to sucrose	As sucrose	None	Typical carbohydrate	Fair
50% inverted	1.0					
100%	1.1					
Fructose	1.3	Slightly thin	Fast without linger	None	Typical carbohydrate	Good
Glucose	0.7	Slightly thin	Fast without linger	None	Typical carbohydrate	Fair
Glucose syrup						
42 DE	0.33		Fast, some linger	None	More than other carbohydrate	Fair
63 DE	0.50		Fast, some linger	None	More than other carbohydrate	Fair
Iso-glucose	1.0		Fast without linger	None	Typical carbohydrate	Fair
Intense sweetness						
Saccharin	350	Slightly chemical sweetness	Slower and persistent	Bitter/metallic aftertaste	Thin	Nil
Cyclamate	33	Slightly chemical sweeteners	Slower and lingering	Of taste at high concentrates	Good	Good
1:10 saccharin/ cyclamate	100	Sugar like	As sucrose	None	Good	Good
Aspartame	140–200	Sugar like	Slight delay, slight linger	Little	Fair to good	Good
Sucralose	450	Sugar like	Fast, slight linger	None	Thin	Nil
Stevia extract	150	Clean sweetness	As sucrose	Slight liquorice menthol/ aftertaste	Thin	Good
Alitame	2000	Clean sweetness	Slight delay, slight linger	None	Thin	Nil
Acesulphame K	100	Good quality sweetness	Fast, slight linger	Bitter at high concentrates	Thin	Nil

a very similar way to the modus operandi of the flavourist. These materials vary dramatically in functional and taste characteristics as well as bulk, stability and interactivity. The principal components used in beverage composition are given in Table 6.14. Different raw materials from the list in Table 6.14 can be used to give similar effects. For example, body, bulk, or mouthfeel are contributed by:

(a) alcohol
(b) water
(c) sugars
(d) fruit or other characterising ingredient, e.g. milk
(e) artificial sweeteners
(f) emulsifiers and stabilisers
(g) acidity regulators

A further investigation would reveal that each of these contribute to the body of the beverage in different ways and it is the expertise of the product formulator to know how the materials interact and to use those most effective for a particular application. Product formulators have to work to very tightly controlled briefs which specify individual requirements including total raw material cost and this has to be considered from the outset of the formulation of a product since it is much more difficult to incorporate such factors at a later stage. To some product formulators this approach is unacceptable since the most difficult stage in the development of any new product is preparing the initial product concepts so that meaningful consumer research may follow. Some would argue that the imposition of precise cost limits prevent these initial steps being carried out correctly when the remainder of the product formulation path will follow smoothly.

An example of the diversity of just one category of principal components used in the formulation of beverages is 'sugars' or, more correctly, 'sugars and artificial, or intense, sweeteners'. The formulator will consider both together as they contribute to the total sweetness of the product. This diversity of character and contribution of sweetness (not to mention intensity) is demonstrated in Table 6.15. A similar list is available for each of the other principal ingredients.

6.6.3 Label Claims

Specific claims which are to be made for the beverage and included on the label must be considered at this juncture, e.g. is the product to be low calorie, is it to offer vitamins, or contain only natural flavourings, etc.? If not taken into account at an early stage, label claims may be difficult to support after the product is formulated.

6.6.4   *Sweetness/Acid Ratio*

This is a very important product attribute which, in many ways, controls the basic characteristics of the beverage in conjunction with the major ingredient/s. We have already seen how the sweetness/acid ratio, or 'balance', in ginger ales can dramatically affect the perception of the added flavouring in dimensions other than sweetness/sharpness alone. A flavouring may be acceptable at one ratio and unacceptable (i.e. without modification) at another or acceptable with one sweetener system and unacceptable with another (e.g. sugar and saccharin). An example of this is the formulation of carbonated lemonade which contains no fruit and uses the harshness of saccharin in its sugar/saccharin sweetener system (medium total sweetness/acid ratio) to give a brighter, harsher more tangy edge than a beverage utilising a terpeneless lemon essence (see Table 6.4). Without the effect of saccharin this flavour would be much 'rounder' and 'sweeter' tasting. A typical carbonated lemonade formulation is given in Table 6.16. An exercise often carried out in the training of a beverage formulator is to attempt to achieve an acceptable tasting carbonated lemonade without using added flavourings.

Table 6.16   Sparkling lemonade (carbonated beverage): materials and quantities for 100 litres finished product

Material	Quantity	Units
Granulated sugar	60.00	kg
Sodium saccharin dihydrate	0.088	kg
Sodium benzoate	0.175	kg
Citric acid anhydrous	2.28	kg
Sodium citrate BP	0.38	kg
Lemon essence terpeneless	0.9625	litres
Water to 1000 litres	~961	litres
Carbon dioxide to specification		
Total	1000	litres

Directions/special notes
Add ingredients in the order listed, dissolving solid ingredients in water before addition, to make a suitable syrup; after adequate mixing, use this syrup for bottling diluting one volume of syrup with an appropriate volume of carbonated water; fill into suitable bottles at the correct vacuity, adequately rinse bottle threads and sealing surface; close and label appropriately

Analytical characteristics

Refractometric Brix (% w/w at 20°C)	6.0	± 0.3
Acidity as citric acid monohydrate (% w/v)	0.25	± 0.02
Sodium saccharin dihydrate (% w/w at 20°C)	0.0088	± 0.00003

Calculated sweetness

Due to sugar (% w/v)	6.00
Due to saccharin (dihydrate) (% w/v)	4.40
Total	10.40

Table 6.17   Manufacture (distillation) of gin

---

(1) The distillation is undertaken at atmospheric pressure in special vessels; all the solid
ingredients are finely divided prior to addition (see separate section)

8.25 kg	juniper berries (mixture of different origins including Italy, Yugoslavia and Russia)
0.875 kg	coriander
0.85 kg	angelica root
1.125 kg	bitter orange peels (of mixed origin)
0.35 kg	sweet orange peels
155.0 g	orris root
100.0 g	liquorice root
35.0 g	cassia
35.0 g	cardamom
35.0 g	calamus root
975.0 kg	alcohol, 95%
1200.0 kg	water

and distill to give first fraction of
975.0 kg   gin distillate

(2) The second fraction is redistilled and used in the next batch

---

6.6.5   Alcoholic Components

Alcoholic components in beverages must be considered concurrently with
the sweetness/acid ratio and sweetness type since they are interactive. All
alcoholic components provide body and bite but vary dramatically in other
contributions, e.g. flavour. Alcohol may be contributed in the form of almost
flavourless spirit of 40% alcohol through strongly flavoured red, white and
fruit wines to beers (including lagers) ciders and their low alcohol varieties
which all contribute different tastes. Some alcoholic beverages use no
sweetener or acidulant at all and here the flavourist controls the composition
of the whole beverage. An example of this is gin manufacture which is
described in Table 6.17.

From the foregoing it can be seen that by this stage in the formulation of
the beverage the 'category of beverage' as defined in Section 6.2 is decided and
may be used as a guide to choosing the correct flavour extract or extraction
method as well as dispersant or solvent.

6.6.6   Water

Even the choice of water with different analytical parameters must be considered for interactivity with flavourings selected for the beverage. Different
methods of water treatment result in varying levels of different residual
chemicals. The most noteworthy of these is chlorine and its reactivity
(particularly noticeable and damaging to a beverage such as carbonated
lemonade) with lemon oil components producing an unpleasant taste. One of

the methods used to protect such a flavour is to use low levels of sulphur dioxide ($<$ 5 ppm) in the product to 'block' the action of chlorine [5].

The only other consideration here is the effect even these low levels of sulphur dioxide may have on the flavouring (see below). It has also been shown [6] that the chlorination of water with sodium hypochlorite sterilants can cause off-flavours in orange flavoured beverages prepared from fruit concentrates (containing low levels of orange oil) preserved with sulphur dioxide. It is believed that this off-flavour is caused mainly by sodium chlorate and other impurities in the hypochlorite and this method of sterilisation should be avoided.

This effect of chlorine in water supplies can also be seen at the consumer end of the retail chain with the preparation of tea and coffee prepared with the new generation of plastic jug kettles. The constant re-boiling of the same quality water with appreciable levels of chlorine causes interactions with phenolic components in the material used in the construction of the kettle to produce chloro-phenol off-flavours in the products made using the boiled water. This off-note has low awareness thresholds in many people.

6.6.7   Characterising Ingredients

It can be seen from Table 6.16 that the main characterising ingredients in non-alcoholic and some alcoholic beverages are usually:

(a) fruit
(b) extract of natural plant material, etc.
(c) milk/yoghurt
(d) other, e.g. glucose syrup

We have already seen for example how milk can provide both body to the mouthfeel and a less acidic taste in beverages, with the examination of the strawberry milk beverage.

6.6.7.1 *Formulating with fruit.*   It is for these reasons that the beverage formulator must consider the effect and impact a particular characterising ingredient will have on the product composition and performance. This may be directly, e.g. the taste and body contribution of fruit juices, or indirectly via the process requirements the use of this ingredient necessitates, e.g. the pasteurisation and preservation requirements of beverages contain fruit. This is best exemplified via the formulation of a 'comminuted' or 'whole' fruit drink and a formulation for such a product in an orange flavour is given in Table 6.18. The legal requirements for such a product in the United Kingdom (at the time of going to press) [7] are:

(a) minimum 10% w/v potable fruit
(b) minimum 22.5% w/v added carbohydrate

Table 6.18 Orange squash: materials and quantities for 1000 litres finished product

Material	Quantity	Units
Granulated sugar	340.88	kg
3:1 Orange base blend	60.00	litres
4.5:1 Orange juice concentrate	19.55	litres
Sodium benzoate	0.296	kg
Sodium metabisulphite	0.226	kg
Citric acid anhydrous	15.36	kg
Tri-potassium citrate	2.00	kg
Ascorbic acid (to give 18 mg/100 g)	~ 0.16	kg
Orange essence		
Water to 1000 litres	698.00	litres
Total	1000.00	litres

Directions/special notes
This product must not be held as a syrup once it has been prepared, without pasteurisation

Analytical characteristics

Refractometric Brix (% w/w at 20°C)	34.0 ± 0.5
Acidity as citric acid monohydrate (% w/v)	2.00 ± 0.05

Calculated sweetness

Due to sugar (% w/v)	34.1

    (c) maximum for artificial sweeteners
    (d) maximum for preservatives
    (e) controls on many other additives

The requirement for 'comminuted' as 'whole' fruit means that a certain level of oil-soluble materials will be dispersed throughout the ingredient, e.g. orange oil in orange comminute and this will impart an intensity of flavour dis-

Table 6.19 Terpeneless orange 'top-note' essence

Mixture of:

(1)	113.00 kg	orange oil[a], cold pressed concentrate 10-fold by vacuum distillation with removed of terpenes to give
	11.30 kg	10-fold orange oil
(2)	11.30 kg	10-fold orange oil
	55.00 kg	alcohol, 95%
	70.00 kg	water

Procedure
Agitate well and allow mixture to stand in separator for 24 h; remove terpenes and waxes to give

190.00 kg	terpeneless orange essence
30.00 kg	alcohol, 95%
100.00 kg	terpeneless orange 'top-note' essence

[a] The flavour taste of the 'top-note' essence may be adjusted by the selection of country and variety of origin of the orange oil.

proportionate to the juice content (% v/v) present. If the product is to be cloudy then, as long as physical stability of the oil is maintained in the comminute and the finished beverage, there will be no need for terpene removal. In this situation, the requirement of the product formulator from the flavouring will be the supplementation of the 'juice' character of the fruit if the orange oil content in the beverage is approximately 0.06% v/v (expressed as single strength oil) and of the 'peel' character of the fruit if the oil content is < 0.01% v/v (expressed as single strength oil). Such a flavouring is known as a 'top-note' and a formula for a suitable essence is included in Table 6.19.

6.6.8   Other Ingredients

6.6.8.1   *Vitamins and nutrients.*   If vitamins and nutrients are to be included in the beverage formulation the stability of these materials and their reactivity must be evaluated under all relevant storage conditions prior to application of the flavourings with which they will almost certainly interact. The B group of vitamins is a good example of unstable composition and off-flavour development in beverages.

6.6.8.2   *Anti-oxidants.*   This is one of the most important areas where the flavourist and beverage formulator must work together to achieve flavour/taste/stability in the finished beverage. If working separately, the flavourist may consider only the anti-oxidant requirements of the flavouring and the formulator may only consider the requirements of the beverage. Colourings, particularly carotenoids, flavourings and some vitamins are susceptible to oxidation both initially and during storage. Oxygen from the atmosphere is entrained in the flavouring materials during initial extraction, e.g. lemon oil from lemons and in the beverage during manufacture, in the headspace of the pack and during ullaging (if multi-use pack) by the consumer.

Oxygen can also diffuse through partially permeable plastic barriers used in packaging materials for both flavourings and beverages. Oxygen can lead to the breakdown of organic molecules in the flavouring and in the beverage and is often accelerated by sunlight and heat. Citrus oils in flavourings and fruit beverages are most susceptible to oxidation. The most common protection for the oil is the use of BHA, BHT and tocopherols.

Removal of terpenes reduces the susceptibility of the citrus oils to oxidation and flavourings based on terpeneless citrus (or 'folded') oils exhibit better stability. Such flavourings will require the expertise of the beverage formulator to provide some of the flavour characteristics of the terpenes by other routes. Since this is extremely difficult, the alternative available to the formulator is to provide and maintain a reducing environment in the beverage in which the terpenes will be more stable. Ascorbic acid and sulphur dioxide may be used for this purpose but the latter causes interactive problems of its own. Ascorbic acid can be introduced into the oil phase as the palmitate.

6.6.8.3 *Preservatives.* Benzoic acid is possibly the most widely used beverage preservative, probably owing to its low taste threshold, low volatility and wide antimicrobial spectrum. For maximum effectiveness of benzoic acid, the pH of the beverage must be about 3.0 or less; benzoate resistant yeasts are widespread in manufacturing plants. Benzoic acid is relatively non-interactive with flavouring materials.

Sorbic acid is used increasingly in beverages and provides good protection through its non-volatility and effectiveness against yeasts. Unfortunately sorbic acid has a lower taste threshold over a wide range and in citrus flavours some consumers are very sensitive to its particular flavour.

Sulphur dioxide is in widespread use as an alternative to benzoic acid and has broad antimicrobial and antiyeast activity in a pH range of 2.0–4.0. It also confers antibrowning and anti-oxidant effects. Generally, sulphur dioxide is extremely reactive chemically and is unacceptable for use with orange flavours and fruit, cans, thiamin (vitamin B_1) and some natural colours.

When the preservative system is selected by the beverage formulator it is essential that consideration be given (via contact with the flavourist) to the effects one or more preservatives will have on the flavouring components so that adequate compensation may be made. This will require sufficient storage testing to reveal longer term oxidative effects as well as immediate reactivity.

6.6.9 *Acidulants and Acidity Regulators*

We have already seen how the effects of the sweetness/acid ratio can be modified by buffering the acid system in the 'American' style ginger ale formulation. The beverage formulator is able further to modify product taste by the selection and use of different acidulants to different effects, e.g. the potentiation of apple taste by the use of malic acid. Certain flavour types react best with certain acidulants, e.g. American cream soda with tartaric acid and it should be remembered that titratable acidity and pH are not the only indicators of the performance of an acidulant. Different acids have different taste effects as an integral part of their character and these can be further modified by the use of buffers.

The best known example of the use of alternative acidulants to citric acid in beverages is cola drinks. Full use is made of the body and sweetness of the sugar content to offset the 'punch' achieved by the use of phosphoric acid. An examination of the cola beverage formula in Table 6.20 shows how the use of a cola nut extract alone is inadequate when used alone for flavouring a beverage. The flavour is 'filled out' by the formulator with either a single cola flavour (as in Table 6.3) or the individual citrus components and spices of that flavour carefully balanced with the other principle components of the beverage. This truly is the work of a flavourist and beverage formulator in harmony, as in Table 6.21.

Table 6.20   Cola carbonated beverage: materials and quantities for 1000 litres finished product

Material	Quantity	Units
Granulated sugar	94.50	kg
Sodium saccharin dihydrate	0.0210	kg
Sodium benzoate	0.1770	kg
Cola extract	6.0000	kg
Bitter principle	0.0200	litres
Cinnamon flavour	0.00075	litres
Lactic acid (edible grade)	0.1870	litres
Cola flavour	0.1500	litres
Citrus flavour	0.0480	litres
Water to 1000 litres	~937.00	litres
Carbon dioxide to specification		
Total	1000.00	litres

Directions/special notes
Add the sugar as a suitable syrup; dissolve the sodium benzoate in water before adding to the batch; after thoroughly mixing, dilute the cola extract with an equal volume of water and add to the batch; the essence and citrus flavour should be blended together before addition, and added to the syrup just before making up to the final volume; make up into a suitable syrup and mix thoroughly; allow the syrup to stand for at least 2 h before commencing canning/bottling

Analytical characteristics

Refractometric Brix (% w/w at 20°C)	9.5 ± 0.5
Acidity as phosphoric (% w/v)	1.106 ± 0.005

Calculated sweetness

Due to sugar (% w/v)	0.945
Due to saccharin (dihydrate) (% w/v)	0.105
	1.050

Table 6.21   Cola essence (for use in conjunction with cola extract)

45.00 kg	terpeneless lemon oil
38.70 kg	terpeneless orange oil
15.00 kg	terpeneless lime oil
1.00 g	cinnamon flavour
0.30 g	nutmeg flavour
100.00 kg	cola essence to be used with cola extract for flavouring beverages

6.7   Summary

A comprehensive review of all beverages and their compositions as they relate to flavourings is not possible in a single chapter. What it has been possible to achieve is a broad overview of beverage categories and the types of flavourings typically in use with methods of extraction, solubilisation and concentration of beverage flavourings. In-depth studies on the use of ginger in beverage formulation and flavouring composition has enabled the reader to gain an insight into the complex and interactive role of the flavourist and applications technologist. An overview of the complexities of beverage formulation has been included with detailed information on the diversity of the choice of sweetness/acid ratio and its interactivity with some perception of flavour. This is one of the most important ingredient areas for the beverage formulator and flavourist to consider when formulating beverages. The interactivity of the flavourist and beverage formulator when the composition of beverages is in discussion is seen to be of the utmost importance.

7 The flavouring of confectionery

D.V. LAWRENCE

7.1 Introduction

The principal ingredient in all confectionery is sugar (sucrose), which in its refined form has little flavour apart from its inherent sweetness. Raw (unrefined) sugar has its own particular flavour, which will be dealt with later in the chapter (see Section 7.3). Other important carbohydrates used in confectionery are corn syrup, invert sugar and dextrose, which are added mainly to control or prevent crystallisation. The texture of the confection may be altered by their use, and this property is used by confectioners to manufacture many varied products.

Other ingredients such as gums, pectin, gelatine, starch, milk, butter, other fats and cocoa, do most to give special textures, although it must not be forgotten that air and water probably have the greatest effect in confectionery. Other ingredients which also play a part include liquorice, honey, nuts, coconut, raw sugar (molasses), malt extract, dried fruit, fruit, and fruit juices. These ingredients are added usually for their flavouring properties, or for their contribution to the eating quality, mouthfeel or nutritional value of a confection. Some products owe their total appeal to these added ingredients. The flavour industry also provides extracts, concentrates and flavourings to suit requirements for all these confectionery types.

Temperature and cooking (or heating) times also play an important role in determining final taste and texture as they have a significant effect on flavour and flavour development.

Figure 7.1 shows temperature bands for producing various confectionery types. The apparently large range is normal, and takes into account recipe differences and texture required. Lower boiling temperatures enable crystallisation to occur and a variation even as small as 0.5°C can make a significant difference to the texture of most types of confectionery.

The principal types of (sugar) confectionery are as follows:

high boilings	cream and lozenge paste
fat boilings	compressed tablets
toffees	jellies and gums
fudge	chewing gum

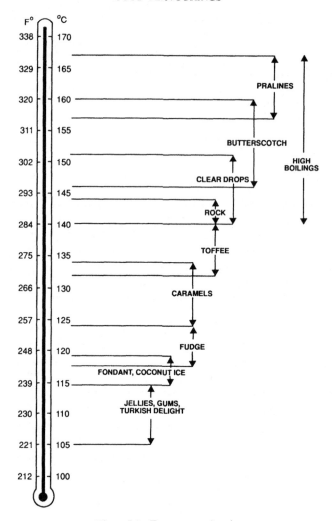

Figure 7.1   Temperature bands.

fondants	panned work
candy	chocolate
paste work	

Typical composition and procedures for the various types are outlined but no account has been taken of water, since it is used mainly to dissolve sugar, or other ingredients or to disperse gums. It is then removed by boiling or drying. The role of flavourings is also discussed.

7.2 Basic Confectionery Types, Recipes, Inherent Flavours

7.2.1 *High Boilings (Hard Candy)*

Candy is a collective U.S. name for sugar confectionery, whereas in the United Kingdom it describes a special crystallised type.

Typical composition:

Sugar	50–70%
Corn syrup 42DE	30–50%
Acid (citric, tartaric, lactic)	0–2.0%
Flavouring	q.s. about 0.1%
Colouring (synthetic)	q.s. about 0.01%

If natural colourings are used, generally many times more is required.

Manufacturing method: Sugar is dissolved in water and corn syrup added. The mixture is boiled to the required temperature, for example 147°C for a 60/40 sugar/corn syrup mix, and cooled. Acid, flavouring and colouring are then added and the resultant material moulded by various means to make the finished confection. In large scale production, liquid corn syrup is metered into sugar solution and cooked in a microfilm cooker (so-called because a thin film of syrup is heated and brought to the required solids content under reduced pressure, in the shortest possible time). Apart from being energy efficient, no browning occurs and therefore little or no cooked flavour is apparent. This syrup is fed into a mixing chamber where calculated quantities of flavouring, colouring and acid are added by means of dosing pumps. The ingredients are then mixed and formed into a ribbon for cooling. The mass is finally transferred onto sizing rollers, prior to spinning to a rope and moulding. On a small scale the batch is boiled in a pan, which may also have a facility to remove final amounts of water by vacuum. At the temperature required, the product is transferred to a cooling table (confectioners 'slab', which has the facility to have hot or cold water passed through it), where the batch is cooled. When the correct temperature is reached, as determined by the viscosity of the mass rather than any other factor, the flavouring, colouring and acid are added, folded in, and the confection finished as before, or passed through 'drop' rollers. 'Drop' here means the shape of the sweet (e.g. pear drops) (see Figure 7.2).

Another type of high boiling (candy) which should be mentioned is the 'deposited' type, where the cooked, flavoured, acidified and coloured syrup is held at high temperature in special hoppers and deposited into metal moulds, prior to cooling and wrapping. The acid used has to be buffered to prevent inversion of sugar, whilst the colouring and flavouring used must be specially selected to withstand the extra heating necessary. Invert sugar is produced by the addition of acid into the boil, or by long slow cooking. Excessive inver-

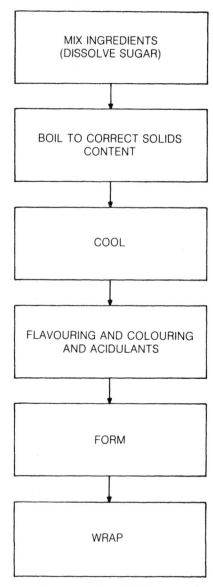

Figure 7.2   Processes used in boiled sweet production.

sion leads to stickiness, or even a product that will not solidify. While some confections have invert sugar added because it controls the crystallisation of super-saturated sucrose solutions, it is usually added as commercially available material or as golden syrup, treacle or honey. Any additional production of invert sugar during cooking needs very careful control and buffer salts are added to adjust the pH and consequently the rate of inversion.

Using the same recipe, 'pulled sugar work' (seaside rock, figured lollypops, satins) can be made. Their manufacture necessitates 'pulling' the previously boiled, flavoured, coloured and acidified sugar/corn syrup mixture on a machine or over a hook, to incorporate air. When moulding (to make fancy shapes or to build up letters) is complete the batch has to be kept hot for a considerable time, and this inevitably leads to deterioration of the flavour. For this reason heat-stable flavourings or lower quality products without fine topnotes, which would be lost, may be utilised. One way of introducing fine flavours into boiled sugar is to prepare delicately flavoured centres. Fruit pulps may be used to produce jams; nut pastes may be prepared, or whole nuts used, as well as all kinds of fillings based on chocolate. They are introduced into the boiled sugar rope by means of a centre pump. The high boiled casing then protects and encloses the lower boiled portion, enabling its finer flavour to be retained.

Very many different confections are made using these same basic ingredients, some boiled to the higher temperature range (e.g. satins) while those boiled to lower temperatures are allowed to crystallise (e.g. Edinburgh rock).

7.2.2   Fat Boilings

Looking at variations on plain boiled sugar as described, the addition of fat is perhaps the most obvious. Traditionally this fat was butter which imparts a smooth mouthfeel, excellent taste, and is self-emulsifying. Butterscotch is made by the addition of 4% or more of butter solids, and the flavour of this product is developed by exposing the raw materials to the high temperature of manufacture.

7.2.2.1   Butterscotch

Typical composition:

Sugar	50–60%
Brown sugar	15–20%
Corn syrup	15–20%
Butter	5–15%
Salt	0.5%
Flavour(s): lemon, vanilla, butter	0.1–0.2%

Manufacturing method: Sugars and corn syrup are dissolved and boiled together until a temperature of about 145°C is reached. Butter is then added and gently incorporated to preserve as much of its flavour as possible. The batch is then boiled to the final temperature of 145–160°C. Where higher temperatures are preferred, special arrangements for direct heating (gas) may have to be made, since they are often too high for steam heated equipment. The mass is then cooled and flavouring incorporated, before the product is cut and wrapped. Generally lemon, usually in the form of lemon oil is added, since it is said to neutralise the greasy effect of fat. Vanilla flavourings are often used to enhance the character of the product and butter flavourings are popular too, as they increase the overall buttery taste. Flavourings of commerce intended for this confection generally contain all these components in carefully balanced amounts.

Variations on butterscotch recipes would be to alter the proportion of white to brown sugar (or the replacement caramelised syrups which are available). If the higher boiling temperatures are used to achieve a special 'cooked' flavour and texture, invert sugar has to be either added or made during production, in replacement for all or part of the corn syrup. The inclusion of invert sugar, in one form or another, results in a much less viscous batch, and higher temperatures are required to reach the same solids content. The amount of butter may be varied or replaced totally or partially with other fats. Butter has natural emulsifying agents present, so if other fats are used, an emulsifier, usually lecithin or glyceryl monostearate (GMS), has to be added in order to ensure proper dispersion of fat through the batch.

7.2.2.2 *Buttermint confectionery.* Incorporation of air, either by means of 'pulling', or the addition of 'frappé', will result in Buttermint types or Mintoes; two typical compositions are as follows.

Typical composition:

Sugar (mix of brown and white)	60%
Corn syrup	30%
Butter or other fat (e.g. hydrogenated palm kernel oil (HPKO))	10%
Salt	0.2–0.5%
Lecithin	q.s.
Peppermint oil	0.1%

Method of manufacture: Sugar syrup and corn syrup are heated to about 130°C, when butter (or other fats with emulsifiers), are added. The mass is reboiled to 138°C, allowed to cool and peppermint oil mixed into the batch which is then pulled to the correct consistency, spun into a rope, formed and wrapped. Frappé is made by beating egg albumen (10%) or gelatine (5%) into previously warmed corn syrup. Both these materials have to be dispersed in minimal amounts of water before addition.

Typical composition:

Sugar (mix of brown and white)	50%

Corn syrup	40%
Fat (HPKO)	4%
Frappé	6%
Salt	0.2–0.5%
Peppermint oil	0.1%
Butter flavour	q.s. about 0.1%
Lecithin	q.s.

Manufacturing method: Sugar and corn syrups are warmed to 140°C. Fat and lecithin are then added and when dispersed, the frappé and peppermint oil are carefully stirred in. The mass is allowed to cool and finished as usual. The heat of the batch will expand air entrapped in the frappé, and care must be taken to avoid its loss by excessive handling. To avoid inconsistent batches it is necessary to control the temperatures used, as well as manufacture of the frappé.

The quality of peppermint oil used in this type of product is important; an oil without harsh top-notes will enhance the smooth character of the confection. This may be a natural American oil (e.g. from *Mentha piperita*) or an oil from China or Brazil (*Mentha arvensis*) which will have been dementholised at source and subsequently rectified, to remove the harsher top-notes and undesirable residues. Alternatively a mixture of the two may be used. Most essential oil and flavour houses have suitable blends to offer. Butter flavourings may also be added, especially when not all the fat used is butter. These are often made with a vanilla background, which accentuates the smooth character of the confection, together with diacetyl and butyric acid. Both these substances are naturally present in butter and increasing the proportion of them boosts the final flavour considerably and replaces processing losses. It should be noted they are available as both natural and synthetic materials. Butter 'esters', i.e. obtained from butter, may also be used. Composite flavourings based on peppermint, vanilla and butter are also available.

As with butterscotch and other confections with 'butter' in their description, it should be remembered that U.K. statutory requirements lay down that 4% butter solids must be present [1].

By using only refined sugar and deodorised fat, fruit flavours and acid can be incorporated to make 'chews'. These are fat boilings prepared without brown sugar and caramelised syrup to achieve a bland product with little background taste that will accept fruit flavour. 'Fruit chews' will contain natural concentrated fruit juices as well as flavouring, in order to comply with U.K. labelling regulations [2]. Essential oil blends are often used to flavour orange, lemon, lime, mandarin, grapefruit, spearmint and peppermint types, while nature identical or artificial flavours are normally utilised for the rest of the range. Permitted matching colourings are also added.

7.2.3   Toffees and Caramels

By the addition of milk solids to confections containing fat, toffees and caramels are produced. It is generally accepted that caramels have a higher proportion of fat and milk solids than toffees, although the difference between toffee and caramel is essentially one of texture and the two types of confection merge into one another without any clear dividing line [3]. Final boiling temperatures range from between 120°C and 150°C, the higher the fat and milk content, the lower the cooking temperature becomes, and in consequence the time for flavour development is also reduced. For this reason corn syrup,

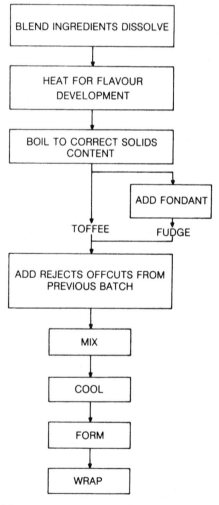

Figure 7.3   Processes used in toffee/caramel/fudge production.

fat and milk are often pre-mixed and heated separately to allow flavour to develop (see Figure 7.3).

Typical composition:

Sugar (mix of brown and white)	40–70%
Sweetened condensed milk	5–30%
(or reconstituted dried milk)	
Corn syrup	20–30%
Fat (HPKO and/or butter)	5–15%
Salt	0.5%
Lecithin	q.s. 1% of the fat content
Flavouring	q.s. about 0.1%

Other ingredients often used are flour, eggs and cream, the last two being obligatory [1] if the product is to be named 'egg and milk' or 'cream' toffee. Malt, treacle, liquorice and chocolate or cocoa can also be added, white nuts and dried fruit can be mixed into the product prior to finishing. Flavourings used are generally vanilla, butter or cream, or mixtures of these, although egg and milk, and other specialised toffee flavours are available.

Method of manufacture: Sugar, with some corn syrup is dissolved in water and heated to about 120°C. Milk, which may be reconstituted dried milk, is heated with fat and the remainder of the corn syrup and mixed to a smooth paste. This mixture is then added to the sugar/corn syrup mix and returned to the boil, 120–150°C depending on the composition. At this point extra ingredients may be added, together with any flavouring, and the mass poured onto a cooling table. It is then cut into suitable sized slabs, scored with toffee cutters and presented as such, or passed to forming rollers, after which the 'rope' is fed to cut-and-wrap machines. Much of this work is not fully automated, with various mixtures being prepared, blended and fed to a microfilm cooker. Flavourings can then be either added via a mixing chamber or by dosing pump directly into the product. The toffee ribbon produced is then fed to a cooling belt and subsequently to forming rollers and finished as described above.

The characteristic flavour of toffee is produced by the heating of milk solids with corn syrup. The higher the temperature and the longer the heating (and cooling) takes, the more flavoured the product becomes. It is therefore most important that the heating processes are carefully controlled to achieve product consistency.

7.2.4 Fudge

It is said that the origin of fudge was 'fudged' toffee where sugars were inadvertently allowed by crystallise [3]. In production, toffee base is made containing a greater proportion of sugar than normal. This is available for crystallisation, and fondant is added to seed the process. Production of fondant is described later in Section 7.2.5.

Typical composition:
Sugar (mix of brown and white)	35–45%
Corn syrup	15–25%
Sweetened condensed milk	10–15%
Fat (HPKO and/or butter)	4–10%
Salt	0.2–0.5%
Fondant	15–25%
Flavouring (vanilla)	0.1%

A similar initial procedure is used to that for toffee, and the mass boiled to 115–118°C. This is allowed to cool in the pan, fondant is added, and the whole thoroughly mixed. Flavouring is then incorporated, and the mass re-mixed. The fudge is then poured into frames lined with waxed paper and allowed to crystallise. The rate of cooling has to be carefully controlled so that the product does not become too coarse or 'marbled' due to uneven formation of crystals. Fudge is usually scored and packed in waxed paper prior to sale.

All the other ingredients mentioned for addition to toffee can be added, together with matching flavourings, to produce special types of fudge, e.g. coffee, walnut, hazelnut, maple and honey. Since the addition of fondant is for the purpose of seeding crystallisation of sugar, any offcuts and other fudge waste can be re-used by addition to the batch at the same time as fondant. Since this recycled material has already been crystallised it will start to seed the batch just as well. If recycled material is used it should be remembered that extra flavour will be produced because the scrap has been heated previously. Good control is essential for product consistency.

7.2.5 Fondant

This product has already been mentioned in the previous section as an ingredient of fudge. However, it also represents a very large part of the confectionery market in its own right. A good description would be 'a controlled crystallisation of sucrose in a syrup medium'.

Typical composition:
Sugar	60–80%
Corn syrup	20–40%

The ratio of corn syrup to sugar is important since the amount of corn syrup added will have a bearing on the final crystal size of the product. Under the same conditions, i.e. boiling, cooling, beating, the more corn syrup present the smaller the crystal size becomes.

The sugar is dissolved in water, taking care that no crystals remain to prematurely seed crystallisation in the batch, and corn syrup added. The batch is boiled to 114–118°C, allowed to cool to 45–50°C, then beaten until crystallisation occurs. This may be done with a wooden spatula on a confectioners slab, or with a special fondant beating machine. During this operation, latent heat of crystallisation is evolved and it is important that the mixing and cooling is continued until this has been dissipated or the crystal size will be increased, resulting in a coarse grained product. The batch is allowed to 'rest' to allow complete crystallisation. The fondant is then re-melted by heating prior to use, or flavoured and coloured, then deposited as described later.

Variation of the recipe, boiling temperature, time of beating and the temperature at which beating starts, all play an important part in determining the final crystal size and consequently the texture. In large scale production, syrups are brought together, boiled in a microfilm cooker, cooled through a heat exchanger, then passed to a continuous fondant beater.

Fondant may be stored and re-melted when required for use in other confections such as coconut ice and fudge or, after the addition of flavouring and colouring, as toppings for cakes and centres for chocolate. In order to make these products, or to prepare fondant creams as such, the mass is deposited into shaped rubber mats or prepared starch trays. This is described in Section 7.2.9. As well as being dipped into chocolate, some may be finished by crystallisation where they are immersed in a supersaturated sugar solution, which, when dried, leaves a hard crust on the outside of the sweet. Fondant can also be used as an enrobing material to make products like maple brazils.

As variations to the above, honey, maple sugar, chocolate, brown sugars (or the equivalent), coffee, milk solids and fat can be added to give different tastes and textures. While peppermint is the most popular flavour, many variations such as almond and vanilla, and florals such as rose and violet, can be used in this confection. The addition of a small amount of acid (about 0.1%) will enhance fruit flavours. Concentrated fruit juices [2] can also be added in addition to the flavouring, while nuts, dried fruit and similar ingredients can be dispersed through the mass. Apart from added ingredients, fondant does not have any inherent flavour and is extremely useful for flavour testing.

7.2.6 Candy

Candy is very similar to fondant, differing only by the size of crystals which are very much larger. This is achieved by starting the crystallisation process at a much higher temperature (as soon as the batch has been brought to the required temperature) and thereby allowing it to proceed for a longer time.

Typical composition:
Sugar   80%
Corn syrup   20%

Sugar is dissolved in water, corn syrup added, and the batch heated to 120°C. It is then allowed to cool slightly, after which the syrup is agitated against the side of the pan with a spatula until it thickens and clouds. Colouring and flavourings are then added, mixed, and the mass poured on to a warm oiled slab to crystallise. It is usually marked as soon as set, and when cold broken into pieces and packed.

Candy can be varied with additions in exactly the same way as fondant, a very well-known type being 'cough candy'. In this product, liquorice and molasses are added together with flavourings based on aniseed oil, clove oil, eucalyptus oil, peppermint oil and menthol. These flavourings are often added for their physiological effect, and many confections of this type do claim to have a beneficial effect. They can be formulated to contain the same amount of these constituents as some medicated syrups.

7.2.7 Cream and Lozenge Paste

These products are variations of fondant, but rather than using crystallisation techniques, finely ground sugar is dispersed in a gum or other colloidal matrix. Fat may also be added, especially when the product is to be used for biscuit cream.

Typical composition:

Ground sugar	50–95%	
Fine caster sugar	0–25%	
Corn syrup	0–25%	
Gelatine or	1%	
Gum arabic or substitute or	4%	or
Gum tragacanth	0.5%	mixtures
Fat (HPKO)	0–10%	
Flavour	q.s.	

Pastes as such can be made by hand by mixing the sugar into gum/gelatine solutions, rolling out into thin sheets and cutting into shapes, and/or sandwiching different coloured sheets together and cutting into suitably sized squares or oblongs, then allowing to dry. In production, large trough mixers are generally used. The mass is then passed through sizing rollers, cut into shapes and normally air dried prior to packaging. Lozenges generally contain only ground sugar, gum and flavouring and are 'stoved' for 24 h, i.e. placed on trays in warm (30–35°C), well-ventilated rooms to reduce excess moisture to around 3%. While the high sugar concentration and gum help to bind the flavourings to the mass, there is still a great deal of flavour loss, which is usually compensated by adding extra amounts of flavouring and careful choice of the flavourings used.

This type of confection includes 'allsorts', 'hundreds and thousands' and similar products. Where fruit flavours are used, acid should be added. Many types of mints and medicated lozenges are made in this way, including 'cachous' which are typically flavoured with 'floral ottos', such as rose and violet. These were originally used as mouth fresheners.

7.2.8   Compressed Tablets

This is a different method of producing lozenges, in which a very high compression is used to give a really smooth texture. This can be varied depending on the particle size of the sugar used.

Typical composition:
Freshly ground sugar	95–98%
Gum	1–3%
Stearic acid	q.s. approx. 1%
Peppermint oil	9.25–0.75%
Water	q.s. to disperse gum and 'wet' batch

Gum is mixed dry with sugar and water then added. Alternatively part of the gum may be added dry, and some as a mucilage (a dispersion of gum in water, usually prepared by overnight soaking.) This mass is then granulated (reduced to small free-flowing particles) and dried. Stearic acid, which is added to act as a lubricant for the tabletting machine dies, may be pre-dissolved in iso-propyl alcohol, together with peppermint oil. This mixture is then added and the whole batch re-mixed and tabletted. For fruit flavours, citric acid should be added to the mix prior to granulation.

If effervescent tablets are required, sodium bicarbonate is added prior to drying, followed by citric acid in the form of fine crystals after removal of water. If liquid flavourings are used they should not contain any water which would activate the bicarbonate/acid mix. Dry flavourings are thus often used in these products. Being encapsulated, dry flavourings have the additional advantage of being non-volatile whilst dry and therefore have a longer shelf-life. Dextrose compresses well without the need of added gums, so no granulation is required and totally dry mixes can be made. Although dextrose is not so sweet as sucrose it makes a very acceptable tablet and dry flavourings are therefore essential.

7.2.9   Jellies and Gums

Gums were originally made with gum arabic, and while it is still used, many substitutes are now available which may partially or totally replace it. Examples of these alternatives are gelatine, agar, pectin, starch and modified starch. Different choices and mixtures of these ingredients will give textures ranging from hard gums, pastilles, soft jellies to turkish delight and marshmallow. The latter product is made by beating to incorporate air to achieve the desired effect. The flavourist should remember that because gum arabic, gelatine and starch require a large amount of water for dispersion, it is not possible to evaporate the excess by boiling in the usual way, since the mass becomes too viscous. Excess water is therefore removed by heating the sweets in starch moulds, sometimes for several days, in order to reach the sugar concentration required to make a satisfactory product. This process, called 'stoving', is carried out by placing the starch filled trays containing the confection in rooms kept at elevated temperatures (e.g. 35–50°C) with a flow of dry air. This gives the typical tougher external texture, which is so well liked but it also subjects the flavouring to prolonged heating and oxidation. For this reason only very cheap essential oils are used for the citrus flavours, because the finer top-notes in high quality oils would be lost or degraded. Other fruit flavourings need to be strong and able to withstand the high temperatures required. On the other hand, pectin based jellies are not subject to this sort of treatment, and since they are boiled and cast at relatively low temperatures (approx. 106°C), they are perhaps one of the best mediums for very fine flavours (Figure 7.4).

Typical composition:

Sugar	40–60%
Corn syrup	40–60%
Citric acid	0–1.5%
Gelling agent	see below
Flavour	q.s.

Alternative gelling agents are used at the following typical levels:

Gelatine	3–10%
Agar	2–3%
Pectin '100'	1.5–2%
Starch (or modified starch)	5–10%
Gum arabic	40–60%

The figures after pectin denote the grade, and 100 means that 1 part pectin will gel 100 parts sugar syrup. Because starch requires prolonged heating and so much water to disperse, pre-gelatinised starches are available, which are in effect pre-cooked or chemically modified. They are used to advantage in gums either wholly or partly in place of gum arabic. They are also used in turkish delight to reduce cooking times.

Typically, the chosen gelling agent is dispersed in water (sometimes overnight soaking is required) and heated to boiling point. Sugar and corn syrup are added and the batch re-heated to boiling point. The final temperature will depend on the type of jelly, but 106–107°C would be typical

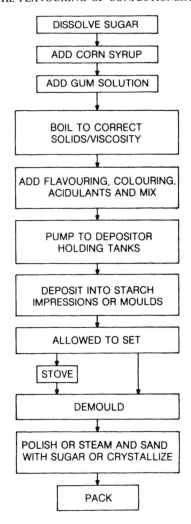

Figure 7.4 Processes used in jelly/gum production.

for most products. Where the mass becomes too viscous to cast at temperatures lower than this, 'stoving' is required to remove excess water. Pectin based products rely on low pH to set, so acid is added just prior to pouring into moulds. If the acid levels, which are required for flavouring as well as completing the set, are too high, buffer salts are added for control (a pH of 3.3 is the usual target). Flavourings and colourings are then added and the syrup deposited into rubber mats or impressions in starch. After setting, sweets so produced are de-moulded and are oiled, cast into sugar or crystallised to prevent them adhering.

In large scale production, this whole operation revolves round a starch moulding plant. Trays are filled with starch, which receives impressions of the sweets to be made from rotary dies. These are then conveyed under depositing heads where product is metered into the impressions. It is usual to have hot syrup pumped to tanks above the depositor, where each tank has the requisite flavouring, colouring and acid added. By this means up to six different flavours and colours can normally be handled at the same time. After depositing, the filled trays are dusted with starch and put aside to set, or taken to the drying rooms. When the correct water content has been reached, trays are returned to the plant where the confections are de-moulded and sent for finishing. A percentage of the starch is automatically removed after each cycle for drying, and returned for refilling into the trays.

7.2.10 Chewing Gum

This product was originally based on natural gum chicle, which is similar to gutta-percha and is the dried latex of *Sapodilla* trees. Supplies of this were insufficient to meet the demands of chewing gum manufacturers, and in consequence substitutes have been developed. These are said to be a blend of resins, oils and rubber, the manufacture and use of which have been covered by patents, mostly of U.S. origin [3]. Chewing gum is a mixture of gum base and corn syrup, which is thickened with fine ground sugar.

Typical composition:

Corn syrup	40–45%
Ground sugar	40–50%
Gum base	10–15%
Stearates or beeswax	1–3%
Flavour (oil blend)	0.25–0.5%

Manufacturing method. Corn syrup is heated to 124°C and the shredded gum base added and mixed until an even distribution of gum is evident. Ground sugar is placed in another mixer and the gum/syrup blend mixed while still warm when flavouring is added. The product is very sticky and liberal applications of dusting powder, usually consisting of ground sugar and starch, are required to handle the mass. Chewing gum is finished in the same way as described for cream and lozenge paste, i.e. rolled out into thin sheets, marked, cooled and broken. Commercially, very large mixers are used for this product, all with the facility to be heated and cooled, and finishing is carried out by sizing rollers, cutters, cooling tunnels and specialised wrapping machines.

It should be noted that the flavouring plays an extra role in this confection. Not only does it provide the taste but it adds to the pliability of the gum. For this reason essential oil blends or specially formulated flavourings are used.

Peppermint and spearmint oils are particularly suitable, as are flavourings with high ester contents (e.g. pear and banana). On no account should these flavourings contain any water, or solvents such as propylene glycol and glycerine. Additional water or humectant would upset the carefully formulated ratio, causing stickiness and result in handling problems.

7.2.11 Panned Work

These products are so-called because manufacture is carried out in large revolving pans. There are two types.

7.2.11.1 Hard panning.

The pans are heated and have a facility to have air blown into them. All kinds of centres are used for this confection, nuts (almonds specially), various seeds, gum, chocolate, boiled sugar, even granulated sugar (used for non-pareils).

The centres are dried if necessary, by rolling in heated pans, or on trays in drying rooms and, if necessary, again sealed with gum solution. This is done by wetting the centres with gum solution and drying. This action prevents any moisture or oil migrating through the layers of sugar which will be deposited on the centres, while the product is in storage.

Centres are placed in the revolving pans and concentrated sugar syrup added, heat and air are applied and moisture driven off. In this way the sugar forms a fine layer on the centre, which can be built up by further applications of syrup. After each syruping, the goods are allowed to fully dry (dusting). Flavourings are added to the pan in small amounts throughout the process, and are absorbed into the coating. Colourings are usually added by dissolving them into the syrups, often only the final coats are coloured, although by using several different colours throughout the process, a sweet which changes colour as it is eaten can be made. Some confectioners use suspensions in syrup of colouring lakes for this work (permitted synthetic colourings are normally available as the sodium salt which is water soluble, although aluminium salts are made which are insoluble and called lakes). These adhere readily to the coating, and being insoluble in water do not colour the mouth when the sweet is eaten. The goods are then finished by coating with edible shellac, and/or polished with wax. Both operations are normally carried out in pans which are retained specially for this purpose [3].

7.2.11.2 Soft panning.

The process of soft panning is mainly carried out under U.K. ambient temperature conditions. Centres which have been prepared previously, and may be almost any type of confectionery including jellies, gums, high boiled products and compressed work, are placed in the pan. Syrups containing corn syrup and sugar in varying proportions (the ratio depends on the type of centre) are used to wet the centres, and fine caster sugar is added. The goods are rolled and further sugar added until the centres are dry. This process of syruping and drying is continued until the confections reach the size required. Finer sugar is used for the final coats to give a smoother finish. Flavourings may be added to the syrups to give an overall effect, although often the centres are highly dosed to give flavour impact when the sweet is chewed, and only the final coats have flavouring added. Quite often this flavouring will be vanilla. Colouring is also added to the final coats and similar solutions or dispersions are used as for hard panning. Flavouring is added, and the goods are glazed and polished in the same way as described for hard panned work.

It is likely that flavouring permeates through the sweets on storage, giving an equalising effect.

7.2.12 *Chocolate*

Considerably simplifying the process, chocolate is made by intimately mixing previously fermented and roasted ground cocoa beans with finely ground sugar, using cocoa butter or other fats as a diluent. Milk chocolate also has milk solids or 'milk crumb' added.

The flavour is dependent on several factors, not the least being the source of the beans. Whilst originating in South America, cocoa is now grown in many tropical regions, one of the most important being West Africa. The beans as grown in pods containing 20–40 almost white coloured seeds surrounded by pulpy flesh. In order to remove this pulp the seeds are carefully fermented for several days. In this process, heat is generated, the germ or embryo is killed, and the initial flavour and brownish colour developed. The beans are then separated from the remains of the pulp and dried in various ways, including drying in the sun [3]. Suitable mixtures from different sources of the fermented beans are made by the user. These take account of the flavour and the market value. Any extraneous matter is removed and roasting takes place in a revolving drum normally heated by hot air at temperatures between 130°C and 150°C where the special roasted flavour is developed. After cooling, the germs and husks are removed and the beans broken into small pieces (now called 'nibs'). These nibs are then ground to produce 'neat work'. This material is ideal for addition to products which require a chocolate flavour. It is rich in cocoa butter some of which may be removed by expression leaving cocoa. This in turn may be converted to 'soluble' cocoa if treated with alkali. Extra cocoa butter is required to increase the fluidity of chocolate, although many substitutes, based on hydrogenated oils are available. These do not have the natural flavour of cocoa butter. The mixture of 'neat work', fat, and fine sugar is refined to make plain chocolate, or with milk solids added, milk chocolate (see Figure 7.5).

   Typical compositions of plain chocolate:
      Ground cocoa nib   35–50%
      Ground sugar   32–48%

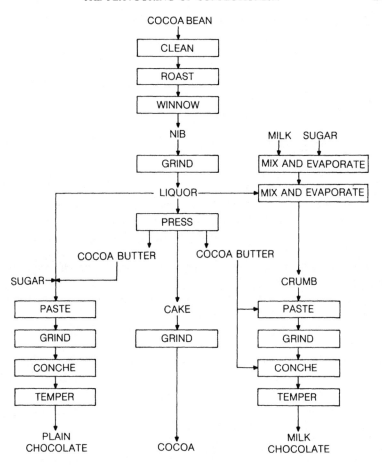

Figure 7.5 Factors affecting flavour in chocolate production (by kind permission of Cadbury Ltd.).

Cocoa butter	5–15%
Salt	0.2–0.5%
Flavour (vanillin)	0.01–0.02%
Lecithin	q.s.

Other fats can be used to replace all or part of the cocoa butter.

Typical composition of milk chocolate:

Ground cocoa nib	10–40%
Ground sugar	20–40%
Cocoa butter	5–20%
Milk powder	10–25%
Salt	0.2–0.5%

Flavour (vanillin)   0.01–0.02%
Lecithin   q.s.

Other fats can be used to replace all or part of the cocoa butter. Lecithin is added to increase the fluidity of the mass. Sometimes milk crumb is used instead of milk powder, cocoa and sugar. A typical composition of milk crumb is as follows:

Sugar   56%
Cocoa mass   7%
Full cream milk solids   37%

Here the sugar and the cocoa mass is dissolved or dispersed in milk before drying and by this means a caramelised, cooked milk taste is imparted, which is characteristic of some brands. Using this material, milk chocolate can be made by merely adding fat, salt and flavouring.

All these operations play an important part in the ultimate flavour. The flavour is also modified by the further treatment of the chocolate mixture (melanger and conche), where the particle size is reduced over a period of several hours and the flavour is modified by driving off more volatile off-flavours.

While vanilla beans can be incorporated into the chocolate mass, their cost now is prohibitive, and either vanillin or ethyl vanillin are more commonly used. Apart from orange, where oil blends are used, very few other flavours are added to chocolate itself, additional tastes being added as flavoured chips or centres. It should also be noted that no water whatsoever is used in chocolate and its presence will completely spoil the setting properties, so flavourings used must be soluble in fat.

Chocolate is re-heated, melted and 'tempered' in order to enrobe fondants and other suitable confections to be used as centres.

7.3 Flavours from Ingredients

Many of these are self-explanatory such as the addition of honey in products like nougat, and liquorice in the many confections containing this ingredient. What may not be so well known in these days of continuous production methods is that, while various grades of brown sugar are readily available (demerara, forths, primrose, dark pieces), they are also sold in liquid form. As well as this, mixtures of the foregoing with refined sugar, in any agreed ratio, can also be obtained. Once the ratio to suit the confection has been agreed, liquid sugar can be delivered by tanker to storage tanks, ready to be pumped to the cookers. This eliminates the weighing and dissolving part of the operation. Most of these liquid sugars are sold at 67° Brix although the high invert products go as high as 81° Brix. The sugar refiners also sell various grades of treacle which is used to advantage in toffee and liquorice confectionery.

It is the effect of heat on these ingredients which is responsible for many of the well-known flavours of products like butterscotch and caramel. If heating is reduced, for example by the use of vacuum and/or microfilm cooking, stronger flavoured syrups are available.

7.4 Flavours Developed During Processing

As indicated earlier in the chapter, an underlying flavour of most confections is derived from the heating of sugars and other ingredients. These ingredients undergo the well-known Maillard and Strecker reactions which are well documented [4]. Products of these reactions are generally found in most confectionery processes. Using traditional gas or solid-fuel fired pans, very high surface temperature are common and higher finishing temperatures (up to 170°C) give rise to special flavours, e.g. praline, butterscotch. Steam heating methods, vacuum boiling and microfilm cookers do not show these effects. The sugar industry provides products which produce sought after flavours by heat treatment. In fact by judicious use of these syrups, extremely good effects can be obtained.

In toffee manufacture it is common practice to heat milk and corn syrup together under very rigidly controlled conditions in order to obtain the desired flavour. Some butterscotch, while being produced in steam heated equipment, undergoes a secondary gas heating to increase the temperature, ensuring the correct flavour is developed, before the product is cooled and finished.

Despite this there are some advantages to be had by cooking with steam and vacuum techniques. This is apart from any cost factors. Because of the lower temperatures used, it is possible to produce fruit flavoured boiled sweets and clear mints without any hint of the caramelisation, which may detract from the desired flavour.

7.5 Selection of Flavourings

Apart from obvious limitations where some flavours are not compatible with other ingredients, e.g. water extracts in chocolate, it is relatively simple to flavour most types of confectionery. 'Essential oils' are used widely and are very suitable. They are natural, available on the open market, and from various sources. An exception is in some jelly products, where absolute clarity is required, and a more soluble flavouring would be used. It should be remembered that all flavourings are best added at the lowest temperature possible, to prevent undue volatilisation and degradation. It follows then that if the nature of the confection decrees that flavourings have to be added either at high temperature, or the products need to be held in heated conditions for any length of time, there will be loss of flavour. In the main, finer top-notes are lost or degraded so the flavour houses will recommend flavours or oils which have additional top-notes added and are selected for their overall suitability.

The main types of 'essential oils' in common use for sugar confectionery are as follows:

Aniseed	Lemon
Clove	Lime
Cinnamon	Orange
Cubeb	Peppermint
Eucalyptus	Rose
Ginger	Spearmint
Grapefruit	Tangerine (Mandarin)

As well as these, 'oleoresins' can be utilised either alone or in conjuction with essential oils. They are almost without exception miscible together and suitable blends can be compounded. The main types used are:

Bay	Nutmeg
Capsicum	Rosemary
Cinnamon	Thyme
Ginger	Vanilla
Marjoram	

All these have the advantages of being 'natural'.

Other single chemical substances either produced by physical extraction or by synthesis, are important in confectionery flavouring, and these include:

Camphor	Eucalyptol
Diacetyl	Maltol
Ethyl maltol	Menthol
Ethyl vanillin	Vanillin

Where compounded flavours are used, higher boiling solvents are preferred in order to reduce volatilisation to a minimum. Propylene glycol, diacetin and triacetin are thus preferred to other more volatile solvents.

Some typical formulations to demonstrate a few of the more common flavours generally used for confectionery are as follows (all the following are made in 'parts by weight').

Almond flavour:

Benzaldehyde	10.000
Vanillin	0.010
Cinnamon bark oil	0.005
Isopropyl alcohol	40.000
Propylene glycol	ad 100.000

Butter flavour:

Vanillin	5.000
Ethyl vanillin	2.500
Butyric acid	5.000
Diacetyl	1.500
Citral	0.002
Propylene glycol	ad 100.000

Butterscotch flavour:

Vanillin	4.000
Maltol	1.000
Maple lactone crystals	0.500
Dihydro coumarin	0.750
Diacetyl	0.500
Butyric acid	5.000
Lemon flavour	10.000
Propylene glycol	ad 100.000

Caramel flavour:

Vanillin	5.000
Ethyl vanillin	1.500
Dihydro coumarin	0.500
Gamma nonalactone	0.100
Caramel colour	1.000
Water	15.000
Propylene glycol	ad 100.000

Coconut flavour:

Vanillin	2.000
Ethyl vanillin	1.000
Gamma nonalactone	5.000
Dihydro coumarin	1.000
Benzaldehyde	0.010
Water	15.000
Propylene glycol	ad 100.000

Creamy milk flavour:

Vanillin	2.500
Ethyl vanillin	3.500
Maple lactone	0.250
Ethyl maltol	0.350
Gamma undecalactone	0.200
Gamma nonalactone	1.000
Delta decalactone	0.250

Acetyl methyl carbinol   0.300
Diacetyl   0.700
Butyric acid   0.500
Propylene glycol   ad 100.000

Ginger flavour:
Oleoresin capsicum 4% capsacine   0.200
Oleoresin ginger   1.000
Ginger oil   1.500
Lemon oil   0.250
Isopropyl alcohol   ad 100.000

Raspberry flavour:
Raspberry ketone   5.000
Maltol   0.250
Ionone alpha   0.050
Ionone beta   0.045
Ethyl butyrate   1.500
Isoamyl acetate   0.400
Isobutyl acetate   2.000
Propylene glycol   ad 100.000

Rose flavour:
Rose geranium oil   2.000
Citronellol   0.500
Geraniol   0.750
Phenyl ethyl alcohol   0.250
Isopropyl alcohol   45.000
Propylene glycol   ad 100.000

Strawberry flavour:
Ethyl maltol   3.000
Ethyl butyrate   2.500
Ethyl valerate   0.150
Isoamyl acetate   0.010
Phenyl ethyl acetate   0.020
Benzyl acetate   0.400
Gamma-decalactone   0.250
Anisaldehyde   0.300
Orange oil   0.100
Isoamyl butyrate   0.040
cis-3-Hexenol   0.100
Isopropyl alcohol   10.000
Propylene glycol   ad 100.000

Vanilla flavour:

Maltol	0.250
Dihydro coumarin	0.500
Vanillin	8.000
Ethyl vanillin	2.000
Heliotropin	0.020
Cinnamon bark oil	0.005
Water	20.000
Propylene glycol	ad 100.000

'Winter' flavour:

Aniseed oil	10.000
Peppermint oil	72.000
Menthol	5.000
Eucalyptol	10.000
Clove oil	2.500
Camphor	1.000
	100.000

All the above flavourings are designed to be used at the rate of 0.1–0.2% in the finished product.

8 Flavourings for bakery and general use

D. ASHWOOD

8.1 Ingredients

Bakery products are, in general, based on three major ingredients and a number of minor, but nevertheless extremely important, components.

8.1.1 Flour

Flour is in most cases the major ingredient and is usually derived from wheat. It may be whole wheat as in the case of wholemeal flour, or part of the wheat berry as in white flour. It can be of different grades depending on its protein content: high protein (11% +) for bread making, and low (approx 9%) for cakes and biscuits. In addition, flour can be treated in various ways to increase the amount of damaged starch cells, a factor which in turn increases its water holding power. It is also possible to treat flour with oxidising agents (typically ascorbic acid) either to increase the apparent strength of the protein fraction, or as is the case in the Chorleywood Breadmaking Process, to reduce the time required for fermentation when used in conjunction with mechanical development. It is also a requirement in law [1] that white flour is nutritionally supplemented with calcium and iron plus vitamins. None of these processes have effects likely to cause major flavour problems. Anyone interested in understanding more about flour and its properties should read *Modern Cereal Chemistry* [2] or some of the other major works on the subject.

There are of course other cereals that can be used, rye, oats and maize being the most common. Rye is of particular interest in that it was normal in continental practice to produce a 'rye sour' by a long period of fermentation, which gives the product a particular flavour and helps to improve keeping properties of the bread. It is possible to reproduce this effect and its shelf-life improvement by the use of flavourings and chemical additions. By this means the risk, inherent in long fermentation periods, of the culture of undesirable microbiological organisms is avoided. The practice of sour dough systems is now little used as it is being replaced by flavourings or specially prepared dried sours made under strictly controlled conditions.

8.1.2 *Sugars*

Sugars are the second major ingredient to be found in most bakery formulae. There are many different sugars that can be used.

(a) Sucrose is the most common both in granular form or as a ground powder; it is also available in its partly refined stage as brown sugars, the best known of which is probably demerara.

(b) Molasses or its partly refined stage golden syrup is another sweetening ingredient manufactured as a by-product of sugar manufacture; it is used not only for its sweetening property but as a flavouring in many products.

(c) Dextrose is produced by the acid or enzymic conversion of starch derived from maize or as a by-product of protein extraction from wheat flour. It is typically available either as a component of liquid glucose syrup or as a powder (dextrose monohydrate). The liquid syrup is supplied with different levels of conversion of starch to sugars, which is measured as dextrose equivalent (DE). The syrup is sold within a range of solids (72–84%) depending on its intended end-use.

(d) Invert sugar, is a product of acid or enzyme treatment; in this case the substrate is sucrose. It has many properties similar to glucose and is often used in formulae for its ability to act as a humectant. We must also consider honey in this group; it is used as a flavouring ingredient as well as a sweetener. There are many honey types all of which have different flavours, characterised by the flowers visited by the bee at the time the honey was produced.

(e) Other bulk sweeteners, fructose and polyols such as sorbitol, are used to make products suitable for diabetics as they do not require the human digestive system to provide insulin.

8.1.3 *Fats*

Fats are the third major ingredient. They are derived from animal and vegetable sources and have to undergo several purification steps before being suitable for bakery use. The processes of filtration, colour and flavour removal, fractionation and hydrogenation allow the manufacturer to produce tailor-made fats for the particular application. The very special flavour characteristics of butter that are changed during the baking process must also be considered. These are the target of much research in the flavour industry, the results of which have produced some excellent flavourings which can add a special note to many baked products.

8.1.4 *Liquids*

Liquids are one of the minor but very important ingredients in bakery formulations, and are normally added in one or more of many forms including, egg, milk and water. The prime function of liquid is to bind all the various additions of the formulae, holding them together in the early stages of the baking process. Later, as the temperature rises, a secondary function of the protein fraction of egg coagulates to produce structure. Free liquid then enters the starch grain of the flour and allows gelatinisation to take place, again adding to the structure of the product. The retention of moisture in the finished item is important in its taste sensation when eaten. Lack of moisture can for example make a cake unacceptable, whilst too much of it can make a biscuit equally unpleasant to eat.

8.1.5 Gases

Gases, producing the effect of aeration, are another minor but important ingredient.

Mechanical aeration, although not strictly an added ingredient, can be produced by beating or whisking, and here egg has a very important role in baked products in that it can hold air in its protein structure. Fat of the correct type will entrap air when beaten. Aeration can also be produced by chemical and biochemical components.

Chemical aeration is possible using ammonium carbonate which on heating decomposes to produce ammonia, carbon dioxide and water. Unfortunately, ammonia tends to re-dissolve into any available water in the product and is therefore not acceptable in high moisture products such as cake. It does form a very useful aerating ingredient for biscuits and low moisture items. Sodium bicarbonate upon heating will release some carbon dioxide, however, the reaction can be made to produce more carbon dioxide when used in conjunction with a variety of acids. Although chemical residues remain after both reactions, they are considered acceptable tastes in powder aerated products. The common acids used are:

Tartaric acid
Cream of tartar
Monosodium orthophosphate
Acid sodium pyrophosphate
Calcium hydrogen phosphate
Glucono-delta-lactone

Glucono-delta-lactone is preferred in that it produces the minimum after taste.

Biochemical aeration by the use of yeast, usually the specially cultivated variety (Saccharomyces cerevisiae), is the prime source of biochemical aeration although it is possible to take advantage of the yeast spores floating in the air or found on the surface of fruits. There are obvious risks in using these so-called wild yeasts as they can be a very unreliable source of aeration and so are little used.

Yeast breaks down the available carbohydrates by the use of enzymes in a fermentation process which produces carbon dioxide gas and a large range of other organic chemicals including ethyl alcohol, pyruvic acid and acetaldehyde, some of which can then go on to further reactions. Many of the chemicals produced have flavour and it is this complex combination which gives bread its unique taste.

8.1.6   Other (Minor) Ingredients

8.1.6.1   *Salt.* Salt is an important constituent of most bakery items and often classified as a flavouring. Its function is that of a flavour enhancer as most products without salt have a flat unappetising taste. It is usually added at the rate of 0.2–1.0% of the finished product depending on the degree of sweetness of the formulation.

8.1.6.2   *Colour.* This is often the first clue to what flavours to expect from a product, and as such can be an important addition to any bakery formulae. The present trend towards natural materials limits the range of colours available and many bakery products are now manufactured without colour addition, requiring added flavouring to be capable of instant recognition.

8.1.6.3   *Flavouring.* Some flavours, such as orange, lemon, strawberry and vanilla are acceptable world-wide while others are peculiar to more specific markets. With the increase of international travel, the range of flavouring suitable for bakery products continues to grow in all countries.

8.2   Bakery Products

Figure 8.1 shows the composition of the major groups of baked items, the *x*-axis showing increasing sugar content, the *y*-axis the fat content. It is not intended that this should be a precise guide as to how to formulate a recipe for any particular item, but serves to show the relationships between the various broad groups of products. These broad product catagories are now examined individually.

8.2.1   Bread

Bread is the basic fermented item consisting mainly of flour with small amounts of fat and sugar. It does, however, contain yeast, which by breaking down the carbohydrate content of flour, produces carbon dioxide gas and a range of complex substances which have natural flavouring effects (see

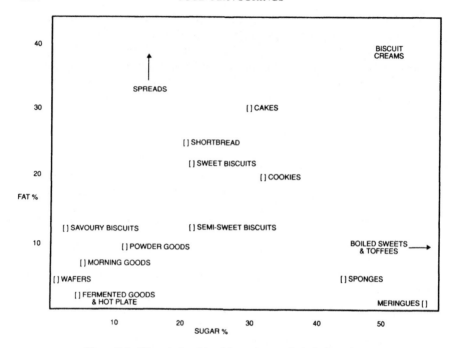

Figure 8.1 The relationship of fat and sugar in baked products.

Section 8.1.5). These flavours become pronounced once they are subjected to the heat of the baking process (see Section 8.2.7). Many attempts have been made to simulate the odour and flavour of baked bread, and some close approximations have been achieved. However, an all-purpose bread flavouring remains one of the flavourists' goals.

8.2.2 Hot Plate Goods

To give a complete picture in this area of low fat and sugar, we must consider products such as scotch pancakes, muffins and pikelets, all of which are baked on a hot plate. This often requires that the article is turned over or cooked on both sides to complete the baking process. These products can be yeast or powder aerated. Wafers are also in this group, and consist of a thin batter baked between two hot plates in a very short time. These are conditions that very few added flavourings can withstand, and therefore, many wafers are bland, tasting only of the flour and possibly milk that has been part of the original batter. They rely on being layered together as sandwiches with biscuit cream or toffee and can be chocolate covered. Vanilla, fruit flavours and nougat are some of the popular flavourings used. Wafers can also be served with ice-cream to provide the flavouring element.

8.2.3   *Morning Goods*

These products are so-called because they are normally made by the baker first thing in the morning following the production of bread. They contain higher levels of fat and sugar than fermented and hot plate goods and are often enriched by the addition of dried fruit and sometimes by the addition of spices such as cinnamon and cardamom. Citrus flavourings are also used, although both citrus oils and spices can, and do, slow down the activity of yeast. This can be compensated for by higher levels of yeast although in the worst cases, yeast activity can be stopped altogether resulting in a dense, unacceptable product. The choice of any flavouring addition must therefore be made very carefully.

8.2.4   *Powder Goods*

These are products that are aerated by the use of baking powder rather than yeast. This category includes scones and raspberry buns, so-called because they contain raspberry jam. These products usually have the characteristic chemical note produced by the aerating agents; they are rarely flavoured with more than vanillin or vanilla flavouring.

8.2.5   *Biscuits*

These products can be subdivided into groups dependent on their fat and sugar content.

Savoury biscuits fall very close to morning goods in terms of fat and low sugar content. Typical flavouring used in such products are cheese and cheese plus products (e.g. pizza); also the use of meat flavouring, spices and additions such as monosodium glutamate with its recognised flavour enhancing ability.

Crackers and laminated pastry (e.g. puff pastry) are made from dough, low in fat and sugar, that has fat laminated into it to produce alternating leaves of dough and fat. During the baking process, steam generated from the dough forces the leaves apart to produce a multi-layer structure. These products are rarely flavoured and are used in the main as carriers for toppings and fillings.

Semi-sweet biscuits can be misleading description in that some of the effect produced by fat is achieved using a chemical reducing agent typically sulphur dioxide added as sodium meta-bisulphite. This has the effect of chemically breaking linkages in the protein matrix that hold the structure rigid and by doing so allow the dough to become extensible and machineable into thin sheets from which biscuits can be made. These chemical conditions make the survival of any added flavouring difficult as they too can be attacked by the reducing agent.

In this product area, the effect of fat adsorption of certain flavour notes starts to become noticeable. This is particularly so in biscuits as they are usually long shelf-life products (8 months shelf-life for a biscuit is normal). During this time, flour tastes begin to reappear and the fat starts to oxidise. This is where a good flavouring can help by suppressing these taints of ageing to an acceptable level.

Sweet biscuits are the major group of biscuits with typical fat content of 17–22% and sugars around 30%. Crisper types can have sugar levels much higher than 30% and ginger nuts are a good example of this. With this higher level of fats and sugars, the baking temperature tends to be lower (e.g. 180°C), and the product can be enhanced by flavouring additions that have quite subtle notes. Vanilla and butter flavouring are used as a basis for many of these product variations.

Shortbread and pastry have higher levels of fat without a corresponding increase in sugar, and in pastry used for savoury fillings no sugar is added. Shortbread and some pastry is made with butter so the use of flavouring, if used at all, is limited to vanilla types so as not to detract from the baked butter notes.

English cookies are so called to distinguish them from the generic use of the word which covers all types of biscuits in America. English cookies have higher levels of fat and sugar than sweet biscuits, and may contain ingredients such as egg, which makes them almost identical in composition to cake, the only difference being lower moisture content (usually below 10%). They can be flavoured more easily than other biscuit types because the baking temperature can be lowered and the higher moisture content retained in the finished article. Many products in this group contain high levels of added ingredients such as chocolate chips, nuts, vine fruits and citrus peel. Part of their sweetening may come from fructose in order to maintain moistness in the finished product.

As the fat and sugar percentage has increased, the technology of manufacture has also had to change. The increases in these ingredients has changed the dough from one capable of being stretched and rolled into a thin sheet, to one that can be rolled into a sheet but not thinly, or alternatively, moulded into a biscuit shape, through to an almost batter-like consistency that requires to be extruded through a nozzle and cut off in an appropriate size piece.

8.2.6 Cakes

Cakes are high in fat and sugar and have finished product moisture content above 10%. As a result they have a shorter shelf-life than biscuits. They invariably contain egg protein as part of their structure and so have this characteristic as a background note to any added flavouring. The baking temperature has again been lowered, and with retention of moisture, flavouring with more pronounced top-notes can be used successfully. It is possible to go even higher in fat and sugar content into the so-called high ratio cake area, whereby the use of specialised fats and flours, very moist sweet cakes can be produced. Like other cake products these can be successfully flavoured if allowance is made for their very sweet taste.

Sponges and meringues are products which do not typically contain fat or oil, although in some sponges small amounts of fat are used. A meringue is egg white and sugar beaten together to produce a stable foam which can be piped out to a suitable shape. This is then cooked at a low temperature which coagulates the protein of the whites and dries out the meringue. They are difficult to flavour because of the long baking time, their low finished water content, and the interaction of the flavouring with the protein content of the product. Sponges on the other hand contain flour and are baked at a higher temperature with the idea of leaving in some moisture. Their high protein content, resulting from the egg, makes them flexible when first baked and they are capable of being rolled into products like swiss rolls after the application of jam and/or cream. The higher moisture content makes flavouring possible, with chocolate being a particular favourite.

8.2.7 *Baking Process*

It will be obvious that some of the above ingredients have flavours of their own that can be enhanced or masked depending on the end result required. To add to this medley of flavours, most bakery goods are subject to a heating process, baking, boiling in water or frying in hot oil which can again add, modify or destroy flavours. These process stages can be very short at high temperatures or long at low temperatures, each giving particular problems in achieving the desired effect.

The process of heating flour-based products brings about complex changes which are still the subject of debate. The increase in temperature of the mass first causes any gas trapped in the structure to expand. This gas can be entrapped air during the mixing process or gas produced chemically as referred to earlier, and at high temperatures, steam from the added water. As the temperature of the mass rises, a point at which the starch starts to gel is reached, and the structure which up to that point is soft and changing, becomes firm and rigid.

The final stage is the onset of browning and the development of the baked flavour notes. This used to be considered as purely the effect of sugar caramelisation; more recent evidence confirms the predominance of the 'Maillard reaction' in the colouring and flavouring of baked products. It is now accepted that many different flavouring chemicals can be produced naturally during the baking process by this reaction. It first produces aroma compounds before the development of colour. As the reaction proceeds, low molecular weight colour and flavour compounds are made, followed in the final stages of the reaction by the production of nitrogenous polymers and copolymers. Condensation of the free amino groups with reducing sugars produce the reaction and it has been shown that the amino acid component has the greatest influence on the resulting flavour compounds produced. The polymerisation and dehydration reactions involved produce complex compounds, which, depending on the temperature, moisture and pH can be extremely varied, Furfural, hydroxymethylfurfural, pyrazines, and short chain aldehydes have all been identified as being produced and research into this area continues (Figure 8.2).

The baking process can be particularly hard on added flavouring. The effect of high temperatures and low moistures in biscuits, for instance, amounts almost to steam distillation of the flavouring with the loss of many of the highly volatile substances and degradation of some of the less volatile components.

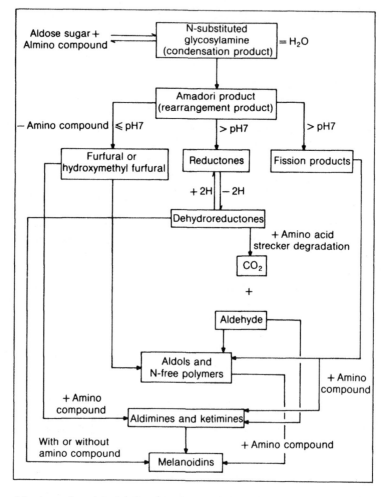

Figure 8.2   An outline of the Maillard reaction. Redrawn by permission of *Food Manufacture* November (1989).

This can be seen most readily in the use of citrus oils in biscuits; the clean notes of the oil become musty and unpleasant. The use of so-called folded oils (terpeneless) which have been concentrated by removing the terpenes, can be used to advantage in this application.

8.3 Bakery Fillings

The various fillings and toppings used with bakery products is an area where flavouring can play an important role.

8.3.1 *Jams and Jellies*

The use of conventional jams and the more specialised agar and pectin jellies has always been an accepted way in which bakery products can be given a flavour after the baking process. Bakery jams often derive their body constituent from sources such as plum or apple. Often, the effect of the named fruit will be achieved by the addition of flavouring to produce strawberry, rasperry and other flavoured jams. Acid conditions exist in most jams and this is particularly true in pectin jellies where it is an essential constituent to bring about the set of the jelly. This acid background will influence the types of flavouring that can be considered, as some flavouring ingredients can be chemically changed in the presence of organic acids.

8.3.2 *Marshmallow*

The original marshmallow was made from an extract of the root of a shrubby herb, *Althoea officinalis*, which produces a gum with a distinct flavour and also the property that when mixed with water, it can be aerated to produce a stable foam. Today, marshmallow is manufactured using egg albumen, milk albumen, and gelatine, or a mixture of these in combination with sugar and added flavouring. The nature of the flavouring used is important in that the foam stability of the materials described is destroyed if small amounts of fat or fat like substances are introduced. It is therefore necessary to use flavouring free from fat and oil. For citrus flavours the use of water extracts from the oil, although weaker in flavour, can produce satisfactory results.

8.3.3 *Creams*

Real dairy cream is not usually flavoured with anything more than vanilla, and until recently even that was strictly not allowed by code of practice. This restriction was agreed by the flour, confectionery and dairy industry to avoid any confusion between dairy cream and imitation creams. Imitation creams are an important area for flavouring additions. They are oil in water emulsions with added emulsifiers, stabilisers and sugar and they have little flavour of their own. Dairy flavourings such as butter, cream and vanilla are required to make them acceptable as a filling and they have advantages in use as they are less likely to support microbial contamination than creams of dairy origin.

8.3.4 Biscuit Creams

Biscuit creams are fat and sugar mixtures made without added water, which would soften the biscuit and make it unacceptable. The use of creams is a convenient way of adding flavour to a biscuit which if added to the biscuit base would be unlikely to survive the baking process. In this way it is possible to produce lemon, strawberry, coffee and many more different flavoured biscuits. However, fat can selectively adsorb flavouring components and can, in extreme cases, totally mask a flavouring during the shelf-life of a biscuit. Figure 8.3 shows the sort of changes a bakery flavouring undergoes from its initial balanced position through mixing, baking, fat adsorption and during the product's life. By recognising and compensating for these changes, the goal of a highly acceptable product can be achieved.

8.3.5 Icings

Icing on the top of bakery goods is yet another way to add flavouring after the baking process. The result of rapidly crystallising sugar and glucose mixtures from saturated solutions produces fondant, a bland material which on warming, flavouring and colouring, can be poured over a product. Variations on fondant can be made by the addition of fat, gelatine and other stabilisers and aerating agents to give icings that remain soft or set hard.

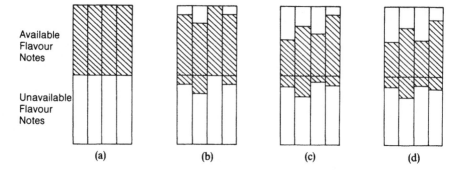

Figure 8.3 Pictorial representation of the changes that take place in a bakery flavouring during processing and product storage. Bars indicate flavour strength: (a) initial balance; (b) balance once included in formulae; (c) balance after baking; (d) balance after storage.

8.4   Summary of Flavouring Characteristics

Some idea of the complexity of flavouring bakery items and the important characteristics of bakery flavouring are summarised below.

(a) It must be capable of withstanding heat processes varying between 100°C and 300°C.

(b) It must be capable of retaining its character in variations of pH between 6 and 8.

(c) It must not be subject to flavour distortion in high fat conditions or must be so formulated to allow for this condition.

(d) It must retain the required flavour impact in low moisture and long storage conditions, such as those occurring in biscuit manufacture or the high moisture conditions of cakes.

(e) It must be compatible with the medium in which it is used (e.g. fat and oil free) when used in meringues and mallows.

(f) It must be cost effective.

9 Dairy flavourings

S. WHITE and G. WHITE

9.1 Introduction

"I hear that you are experts in dairy flavourings. Well, I need a cheese flavouring of general Cheddar type, with a hint of Blue, a sort of buttery, creamy background, and perhaps a slight Swiss fruitiness. I want to use it in a low fat cheese sauce for a low calorie fish and pasta frozen ready meal, designed for both microwave and conventional oven preparation. The flavouring must be a powder with no added flavour enhancers like monosodium glutamate, and it cannot be artificial—in fact the Marketing Department would prefer it to be totally natural. And, by the way, the finished product is to be sold throughout Europe and Scandinavia."

The above transcript of a hypothetical call from a development technologist in a food company is sadly not as typical as the flavourist might hope. All too often the flavour requirements are much less tangibly defined, the end product and processing are not revealed, and the legal and marketing implications of the target market are not clear. This makes the job of the flavourist trying to satisfy customers' flavour needs much more difficult. It is very important that the flavourist engages the customer in dialogue about his or her flavour needs. Only then can the flavourist use broad experience to help to satisfy those needs.

This chapter attempts to categorise the basic areas of knowledge which the experienced dairy flavourist draws upon to understand and fulfil the flavour needs of today's demanding food development technologist. It has been written mainly for the flavourist venturing into dairy flavour types for the first time, but should prove of interest to all interested in other flavourists' views. It is not highly technical, but gives the basics which help to build up a 'feel' for dairy flavours.

You may be surprised by the breadth of information about different types of real cheeses, butters and other dairy products. Such an understanding is vital to the process of defining the flavour target; and the target must be adequately defined if you are to have any chance of reproducing the flavour in a processed food product.

All dairy products start out as milk; their flavour components tend to be similar; the secret of their varied and unique characters is in the balance of those components.

9.1.1   *History of Animal Milks as a Human Food Source*

The earliest domestication of animals is believed to have been about 6500 BC, and with this came the widespread consumption of animal milks by man. It is highly likely that the practice had begun long before this time, with the milk of wild animals hunted for their flesh. The nature of milk itself, with the influence of weather, probably gave rapid rise to a range of dairy products, and this formed the basis of the wide range of milk products we have come to know today. We should be grateful that early man discovered the variety of possible dairy products before he devised refrigeration, a process which might have prevented many of them from developing!

9.1.2   *The Development of Flavour in Dairy Products*

Although there are significant species variations, animal milks are generally oil-in-water emulsions containing varying quantities of triglyceride fats, with the characteristic milk protein casein (plus other proteins at much lower levels), the milk sugar lactose, and a broad range of vitamins and minerals. In short, they contain just the necessary mix of nutrients to ensure the healthy development of the juvenile of the species until it is able to digest other foods.

These macroscopic components are not solely responsible for the varied flavours we associate with dairy products. They provide the raw materials for the development of an immense variety of aromatic compounds. Degradation of the protein, lactose and fat components directly yields many aroma compounds, some more desirable than others! But the great variety in dairy product flavours would not occur without the action of a range of microorganisms which selectively act on the raw milk to give compounds responsible for the familiar flavour characters. Further chemical interaction between these compounds increases the range of chemical species which contribute to dairy flavours.

Differences between the milks of different species' manifest themselves mainly as differences in lactose, protein and fat levels, and in particular in differences in the chemical composition of the fat triglycerides. This in turn gives rise to differences in the types and balance of small aromatic molecules liberated in degradation processes. Thus there is significant variation in flavour between dairy products derived from milk of different species, despite the broadly similar reactions taking place.

9.1.3   *Instrumental Analysis*

As in many areas of flavour science, understanding of dairy product flavours has advanced tremendously in the last 30 years. This has been largely driven by the increasing availability of instrumental analytical techniques which allow detailed examination of the low-level components of complex mixtures. Gas chromatography and mass spectrometry have revolutionised the study of all flavours, and of dairy flavour in particular. Together, these two techniques enable separation and identification of components in extracts from dairy products. However, they do not indicate which components are the most important contributors to the overall flavour of the dairy product. For this the trained human nose is still the most reliable, if not the only means of analysis. Sometimes the nose can usefully be used as an aroma-specific detector for the gas chromatograph, indicating which components should be further studied by mass spectrometry and other techniques. The result can be a list of the key compounds which together characterise the flavour of the product under study. But even this is of little use to the flavourist unless the compounds are available, or could be made available within commercial cost restraints. So, although analytical information is of great help, the skill and experience of the flavourist remains paramount in the development of commercially viable flavourings.

9.1.4 *The Development and Uses of Dairy Flavourings*

Dairy flavourings are used throughout almost all sectors of the manufacturing food industry from snack foods through to alcoholic beverages, from sugar confectionery to ready meals, from dairy products even to specifically non-dairy foods. The range is extremely wide, and this demands a certain approach from the dairy flavourist. It is vital to obtain as much information as possible about the application the customer has in mind. At a minimum, each of the following considerations should be addressed.

End-product:
    type of end-product
    processing
    ingredients—is the recipe available?
    pH of the end-product
    is unflavoured base product available?
    packaging of end-product
    storage of end-product (freeze-thaw?)
    shelf-life required
    countries of final sale—flavour 'culture'
                            —legislation
                            —labelling

Flavouring:
    flavour character required?
    is exclusivity required?
    status required (natural, nature identical, artificial)
    form required (powder, paste, liquid)
    solubility required

strength required (handling problems?)
any solvent or carrier restrictions
packing required

General:
   timescale—urgency?
   workload—priorities?
Commercial:
   target flavouring price
   projected volume
   manufacturing capacity
   costs—flavouring ingredients
      —manufacture
      —packing
      —distribution
   potential earnings—this customer
                     —other customers?
   credit terms
   competition

A flavourist must be a jack-of-all-trades, expert in whatever discipline is required! Intermittently, and concurrently, the skills of chemist, artist, food technologist, designer, marketeer, consumer and even psychoanalyst may be needed in order to identify and satisfy the customer's needs. The customer has an idea of what he wants; the marketing department sets a brief; the food technologists research the practicalities. Ideally, the flavourist works along-side this whole process to achieve the desired flavour profile.

Flavourists can obviously benefit from a detailed knowledge of the compounds responsible for flavour in real dairy products, but this is only a part of the story. The inherent flavour of the product will greatly affect the emphasis of the required flavouring. Many of the components of the 'real thing' will be technically unavailable, or commercially unusable. The flavour of the 'real thing' may not even be what the customer really needs! Before commencing development work on any project, flavourists should satisfy themselves that they fully understand the customer's needs. The best way to do this is to question the customer directly. Such a direct approach is rarely rejected, as most food product developers realise this is the most effective way for them to achieve their goals.

9.2 Milk and Cream

9.2.1 *Whole Cows' Milk*

Cows' milk is by far the most commonly consumed milk in the United Kingdom, so for the purposes of this section, we confine our discussions to cows' milk, and future references to milk should be understood to imply that. Fresh milk is a stable emulsion of fat globules dispersed in a water phase containing both dissolved and suspended non-fat solids. The water content is around 87.5%. It has a slightly sweet taste due to the milk sugar lactose, and this is modified by the fatty acids and their condensation and oxidation products, which are derived from milk fat. The most important fatty acid from the flavourist's viewpoint is butyric acid at 4% of the total fatty acids, with hexanoic and octanoic acids also significantly contributing to flavour.

Many factors affect the flavour of milk, some even before the milk leaves the cow (see Figure 9.1). The cow's diet is of major significance [1], both in the development of desirable flavour and in the responsibility for off-flavours. In particular, chlorophenols introduced into the diet through herbicides, pesticides and disinfectants give rise to well-known taints in milk. Pharmacologically active compounds in the diet may cause physiological changes in the cow which are reflected in the composition and hence flavour of the milk [2].

Even under relatively ideal refrigerated storage conditions, the flavour of milk changes further once it has been obtained from the cow. Oxidation can produce undesirable flavour compounds such as alcohols, alk-2-enals and alk-2,4-dienals. This process even occurs in chilled milk, in which a common oxidation product is undeca-2,4-dienal [3]. Alcohols, lactones, acids and hydrocarbons are also produced by oxidation of unsaturated fatty acids, but not all of these have a negative effect upon flavour.

Exposure of milk to sunlight can cause the breakdown of methionine to methional, which can then react further to produce methanethiol and various methyl sulphides. Heating produces lactones from gamma- and delta-hydroxyacids, themselves derived from the triglyceride milk fat. In cows' milk delta-hydroxyacids predominate, explaining the importance of delta-lactones in milk flavour. The review of Forss [1] gives much more detail of these reactions, and is recommended to those needing to study this area in more detail.

Nearly all milk for drinking sold in the United Kingdom is pasteurised, as is most milk destined for further processing. Although fairly mild, the pasteurisation conditions are sufficient to cause some changes in milk flavour. Most methods of further processing involve some degree of heating and hence oxidation. This effect is sometimes compounded by the addition of other materials, for example sugars, which participate in the development of different flavour characters.

9.2.2   Whole Milk Powder

Whole milk is generally spray dried to form a powdered product. The process has several stages:

    standardisation
    pasteurisation

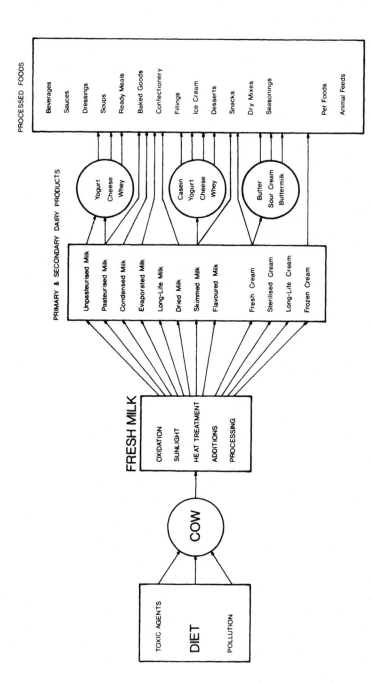

Figure 9.1 Milk products: their source and applications.

evaporation (to concentrate)
homogenisation
drying

Each of these stages affects the flavour of the resulting powder, which has significantly impaired flavour compared to the fresh milk feedstock. The minimum fat content is 26% and this prevents the easy dissolution of whole milk powder. An 'instantised' version can be made by coating the powder with about 0.2% lecithin.

9.2.3 Skimmed Milk

Most fat is removed to produce skimmed milk. It has a much shallower and obviously less fatty taste than whole milk. In recent years increasing consumer awareness of the detrimental health effects of saturated fat in the diet have led to an increasing popularity for skimmed milk. It is widely used in both liquid and especially powder form in food processing. Key uses are in confectionery, ice-cream, dessert mixes, comminuted meats, sauces, animal feeds, and further dairy products such as cottage cheese and yogurt.

9.2.4 Sterilised Milk

Sterilised milk is heated in the bottle to effect sterilisation. Its flavour is definitely that of cooked milk. Key flavour formation occurs through the degradation of sulphur-containing whey proteins to give hydrogen sulphide and other thiols, and Maillard browning reactions with sugars present in the milk. The distinctive flavour shows great regional variations in popularity across the United Kingdom.

9.2.5 UHT Milk

UHT (ultra high temperature) or HTST (high temperature short time) treatment is a process used to effectively sterilise milk so that it can be aseptically packed to achieve long ambient shelf-life. The milk, whole or skimmed, is heated to 122°C (minimum) for one second (minimum). Although a short duration, this high temperature treatment affects flavour significantly, especially in whole milk, where oxidation of the fat occurs. The resulting flavour is much preferable to the distinctly caramel notes of traditionally sterilised and bottled milk, but it is nevertheless very different from that of fresh or normally pasteurised milk. Consumer acceptance of 'long-life' milk has generally been on grounds of convenience rather than taste. The relative absence of fat in skimmed milk results in a UHT product which is much more similar to its fresh counterpart, and this has formed the basis of a whole range of flavoured milk drinks, in which fruit or chocolate flavourings are added prior to the UHT treatment.

9.2.6   Evaporated and Sweetened Condensed Milk

Evaporated and sweetened condensed milks undergo relatively severe heat processing. Not only are they evaporated to remove most of the water content of the milk, but they may also be canned and sterilised at up to 120°C for 10 min. This gives rise to very pronounced cooked flavour characteristics. In sweetened condensed milk, sugar is added before the concentration process, and the resulting flavour is distinctly caramel.

9.2.7   Cream

Cream is the fatty portion separated in the production of skimmed milk. It contains almost all the fat from the milk. Various standardised fat levels are available in the United Kingdom under regulated names: single cream (18%), double cream (48%), whipping cream (35%). The Cream Regulations (1970) [4] regulate the composition and labelling of cream products in the United Kingdom where consumption of cream has been increasing in recent years, but still lags well behind that in other EC member states [5].

9.2.8   Sterilised Cream

Sterilised cream is produced in both cans and jars. Like sterilised milk, the heat treatment causes significant flavour changes and the character of sterilised cream can almost be regarded as a sub-category in its own right. The high fat content ensures that the effects of oxidation are quite marked.

9.2.9   Clotted Cream

In clotted cream, the fat content is increased to 55% by gentle evaporation. This pasteurises the product and forms the characteristic 'crust'. As might be expected from the higher fat content and the further heat treatment, the flavour is stronger than that of fresh cream, but it is more rounded and subtle than sterilised cream.

9.2.10   Casein

Casein (milk protein) can be separated from milk either by the use of acid or rennet to coagulate the protein. It has little flavour and a large proportion is used for non-food purposes. In recent years rennet casein has found important use in the manufacture of cheese analogues, in which the milk fat is replaced by vegetable fat for both economic and health reasons. The other main food use of casein is as the sodium salt, which is used in meat products, coffee whiteners and whipped desserts for its emulsifying, stabilising and water-binding properties [6].

9.2.11 *Whey*

Whey is the by-product of the coagulation and separation of casein from milk. There are four main types:

- (a) sweet cheese whey, largely from cheddar production
- (b) acid cheese whey, largely from cottage cheese
- (c) acid casein whey, from acid casein production
- (d) rennet casein whey, from rennet casein production

Whey powders are widely used in food processing, mainly for their functional properties. However, cheese wheys in particular have more flavour and can be used to provide a cost effective carrier for dairy flavourings. The subtle flavour acts as a useful base upon which to build a realistic dairy flavouring.

9.2.12 *The Applications of Milk and Cream Flavourings*

Milk and cream flavourings find two main uses in food processing:

- (a) products requiring distinct and identifiable dairy characters
- (b) products requiring the 'body/creaminess/richness' which is associated with dairy products.

In fulfilling these two needs, milk and cream flavours find use across the full spectrum of the food industry from animal feedstuffs to sauces, dips and dressings; from baked goods to coffee whiteners [7]; from sugar confectionery to low-fat spreads. They are used for cost savings, to add milk character to non-dairy products, even to provide a stronger overall flavour character than is possible using the 'real thing'!

The most commonly requested flavour types are milk, condensed milk and cream, both in their fresh and sterilised forms. Perhaps the most common requirement is to add 'body' to products formulated with skimmed milk for cost and/or health reasons. This often results in a reduction in palatability compared to the full-fat product. Well-designed dairy flavourings can make a significant difference to such products.

9.2.13 *The Development of Milk and Cream Flavourings*

In approaching the problem of development of this type of flavouring, it is often to best to consider the position(s) of the target character(s) within the spectrum of flavour types ranging from skimmed milk, through whole milk, cream and on to the heated characters of condensed milk. Skimmed milk is sweet, with relatively low flavour impact; whole milk adds some fatty character. Cream is fuller and fattier with lactonic and sulphurous notes; condensed milk has all these characters plus caramel notes resulting from heat treatment.

As in many other areas of flavouring development, there is a tradition of milk and cream flavouring types which bear little relationship to the real foods. They owe their existence to the easy availability of certain flavouring ingredients in the early days of the commercial flavouring industry. In this case these materials are typified by vanillin, maltol, ethyl butyrate and p-methyl acetophenone. Their use in dairy flavourings is more by association than as a reflection of their occurrence, or particularly, level in the foods they purport to emulate. Nevertheless, the flavourist must recognise that there is still a demand for flavourings of this traditional type; not least because the consuming public has come to expect such flavour characters over a fairly long period of time.

Today, however, analytical instrumentation has given the flavourist a much better understanding of the compounds responsible for the flavour characters in real milk and cream. At the same time, many more of these compounds have become commercially available, thereby enabling the development of much more realistic flavourings. Ingredients of key importance in the creation of such realistic flavours include:

delta-lactones	for creaminess
carbonyls	for buttery notes
short chain fatty acids	for cheesiness
sulphur compounds	for heated notes
pyrazines (low levels)	for nutty and green notes
maltol (low levels)	for 'fullness'
vanillin (low levels)	for 'fullness'

Many ingredients in these categories are commercially available in nature identical form. In recent years, in direct response to market demand, many have also become available in natural form, thereby opening up the possibilities for natural milk and cream flavourings.

However, a few characters still cannot be obtained as single natural flavouring substances. For these notes the flavourist must turn to flavour preparations from natural sources. The obvious characteristics are derived from dairy sources, and for example, lipolysed butterfat can be used to impart a useful base character to dairy flavours generally. Many other notes are available from natural vegetable and spice extracts.

In the now limited number of cases where artificial flavourings are acceptable, compounds such as butyl butyryl-lactate can be very useful for imparting a tenacious creamy body.

9.3 Yogurt and Fermented Products

Once ancient civilisations began to domesticate animals and obtain milk from them, the discovery of various means of keeping, storing and processing milk probably followed largely by accident and human curiosity. In ancient times the existence of bacteria was unknown, and many well-known products were the result of unguided empiricism.

Fermentation of milk involves the action of microorganisms; principally the lactic acid bacteria. These sour the milk by converting the milk sugar lactose to lactic acid. Many of these lactic acid bacteria occur naturally in milk, and are responsible for the natural souring process. Since large concentrations of lactic acid are generated, the development of other microorganisms which may cause spoiling, is inhibited, and thus the fermented milk is preserved to some degree.

Action by a range of differing bacillus species result in curd formation in a range of products of varying solubility. In the United Kingdom, the best known example of such fermented milk products is yogurt. However, in other parts of the world, many other types are common, and these form an important source of nutrition to much of the world's population. In this section, yogurt flavour is covered in some detail, together with brief discussions of a representative number of the other product types.

9.3.1   Yogurt

Yogurt is a white semi-solid product with a clean acidic taste and a varying degree of creaminess which depends largely upon the fat content of the milk used in its production. In the United Kingdom, cow's milk yogurt is the most common, but yogurts made with ewes' and goats' milk are growing in popularity, particularly in health food outlets. The source of the milk has a significant effect upon the taste of yogurt produced, and both ewes' and goats' milk varieties have characteristic notes that are not present in the more familiar cows' milk product.

Yogurt is a very versatile foodstuff, which is eaten both as a food in its own right (often flavoured with fruit, sweet brown, vegetable or spice/herb flavourings) or as an ingredient in other foods, both sweet and savoury. It can be used in marinades, dips, sauces, dressings, baked goods, chilled and frozen desserts, to name just a few. Today in the United Kingdom, yogurt represents 64% of the chilled dairy market, which is worth some £600 million per year [8]. The majority of this is fruit flavoured.

The two main bacteria involved in the manufacture of yogurt are *Streptococcus thermophilus* and *Lactobacillus bulgaricus*. Lactic acid is formed, and it is this compound which is largely responsible for the acidic 'bite' and the bacteriological stability. The basic manufacturing procedure is very simple, and can be modelled quite well in the domestic kitchen. Figure 9.2 gives a simplified diagram of the domestic preparation of yogurt. It is perhaps worth mentioning that cleanliness is vital, and it is essential to boil the milk before cooling. A more complete explanation of the processes involved, including commercial practices, is given by Tamime and Deeth [9].

In the United Kingdom, there is no legislation specifically covering yogurt,

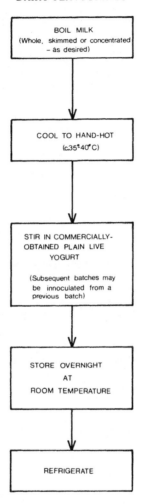

Figure 9.2 Simplified flow-process diagram of the domestic preparation of yogurt.

although the Dairy Trade Federation has published a Code of Practice [10] which covers its composition and labelling.

The flavour of yogurt is produced by the action of the bacteria on milk. Numerous biochemical changes occur during yogurt fermentation, and these are supported by chemical changes, which may involve products of thermal degradation, depending upon the degree of heating used in the manufacturing process. As already indicated, a major flavour contributor is lactic acid, which is derived by microbial fermentation of lactose. A further significant flavour contributor is acetaldehyde. The detailed biochemistry of yogurt flavour development is beyond the scope of this chapter, but more important for the creative flavourist is a broad knowledge of the range of compounds which are responsible for the characteristic flavour of yogurt. However, the review of Tamime and Deeth [9] discusses yogurt biochemistry in detail, and would be a good starting point for those needing to study this area more closely.

Figure 9.3 gives a simplified diagrammatic representation of the flavour

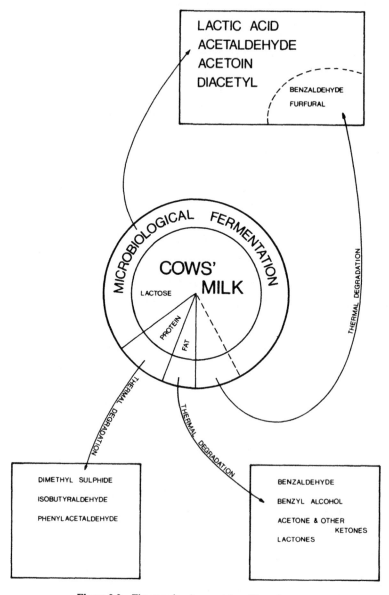

Figure 9.3   Flavour development in milk and cream.

components and their sources in cows' milk yogurt, seen very much from an organoleptic point of view. It is useful to utilise such a diagram when developing a new flavouring, as it helps to clarify the problem and identify the key components of the target flavour. Reference to the qualitative and quantitative lists of volatile compounds in food published by CIVO/TNO [11] reveals over 50 compounds which have been identified in yogurt, and gives quantitative data on 13 of those compounds. However, not all these compounds have an equivalent role in characterising yogurt flavour, so the flavourist must concentrate on those which are:

organoleptically most important
readily available in the required form (natural or nature identical)
cost-effective in use

In yogurt, lactic acid and acetaldehyde are very important, whilst other compounds (both those indicated in Figure 9.3 and others) vary in importance according to the geographical source/style of the yogurt.

So why use a yogurt flavouring? Fresh yogurt is not easily used in products other than wet recipes, and even then it imposes limitations upon keeping quality, generally requiring chilled storage and short shelf-life. Powdered yogurts are available, but although they are generally acidic, they often fail to impart the full aromatic qualities of fresh yogurt. Further, yogurt powders are fairly expensive in use. Finally, and possibly most important, the average consumer's recollection, and therefore expectation, of yogurt flavour may differ markedly from the character of the fresh product. It is this expectation which the food product developer must aim to satisfy. Consequently there is a fairly high demand for high quality yogurt flavourings.

Yogurt flavourings can be used in snack food seasonings, dips, dessert mixes, cream fillings, dressings, soups, sauces and marinades. Consideration must be given to the overall acidity of the product to be flavoured, and the exact characteristics the customer requires. Experience shows that customers often require a degree of creaminess which is not true to real yogurt, and they find a natural acidity level too high for their needs. It is important, therefore, to work with the customer to define the desired end-product profile as completely and accurately as possible.

Undoubtedly the fresh character of acetaldehyde is appropriate in cows' milk yogurt flavour, but in flavouring formulations it must be used with caution as it is unpleasant and dangerous to handle, and has significant labelling implications when used in formulations at 1% or more.

A wide range of nature identical flavouring ingredients are available to the flavourist:

lactic acid   for acidity
lactones   for creaminess

short chain fatty acids	for cheesiness
ketones	for buttery notes
sulphur compounds	for off-milk notes
heterocyclic compounds	for boiled milk notes

Many of the above materials are now also available in natural form, and those which are not can sometimes be substituted with natural flavouring preparations that are rich in the compounds needed.

If the customer's brief extends to the use of artificial materials (which is tending to be less and less common now), the flavourist's palette is extended by a few key compounds which can add a particular tenacity to the dairy character. The best known of these artificial materials is the compound ester butyl butyryl-lactate, the chemical components of which are themselves important dairy nature identicals.

9.3.2   Other Fermented Milk Products

In other parts of the world, many different types of fermented milk product result from the use of different microorganisms. Although the demand for flavourings of these types is small compared to yogurt, they cannot be ignored; and many of them bear little similarity in flavour to yogurt.

9.3.2.1   *Cultured milks.*   Cultured buttermilk is popular in the United States and in Ireland, but is not commonly consumed in the United Kingdom. Buttermilk is the liquid which separates when cream is churned to make butter; it is the traditional source for making cultured buttermilk. However, commercial 'cultured buttermilk' is made by innoculating pasteurised skimmed milk with *Streptococcus cremoris* and/or *Streptococcus lactis*, then incubating at around 20°C to reach a pH of 4.5–4.6. The main flavour contributors are lactic acid and diacetyl, which produce a buttery character with a clean acidic background.

Cultured milks are becoming increasingly popular in the Scandinavian countries. There are several different types, some of which depend on acetaldehyde as well as diacetyl for the major flavour character [12].

9.3.2.2   *Acidophilus milk.*   Acidophilus milk is produced by souring milk with *Lactobacillus acidophilus*, an organism which is often found in the human gut. The product is easy to digest, and has been claimed to have therapeutic effects, supposedly neutralising various disease-producing agents. One of the major flavour contributors is again lactic acid, but the level is not as high as in yogurt, and the flavour is much more like that of sour milk.

9.3.2.3   *Kefir.*   Kefir is a mildly alcoholic soured milk drink, which is generally accepted to have originated in the Soviet Union. It is now very popular throughout the Eastern Bloc and the Middle East.

Kefir is made by the addition of 'Kefir grains' to cows' milk; these are small brownish 'seeds' which are pre-swollen to double their size in warm water before being added to the milk. The resulting chemical reaction sours and ferments the milk. The seeds are dried particles of clotted milk containing cultures of *Lactobacillus brevis*, *Streptococcus lactis* and various lactose fermenting yeasts.

Kefir is slightly effervescent and forms a head rather like beer. It has a creamy consistency and an acidic taste; the lactic acid content is between 0.6% and 0.9% and this is responsible, along with a low level of acetic acid, for a pH of about 4.3–4.4. The alcohol content varies widely between 0.2% and 1.0%. Other important flavour contributors are again acetaldehyde and diacetyl in a ratio of roughly 1:2. In the Soviet Union, fresh Kefir is prescribed to treat gastro-intestinal infections, as it is thought to destroy pathogens [13].

9.3.2.4 *Koumiss.* Koumiss is very similar to Kefir, but is usually made with mares' milk. It tends to have a slightly higher alcohol content of up to 3% because of the higher lactose content of mares' milk. Fermentation is effected by inoculation with a combination of thermophilic lactobacilli and yeasts.

9.3.2.5 *Labneh.* Labneh is concentrated Middle Eastern yogurt, which has the appearance of cream cheese. It is rolled into small balls and stored in olive oil. Regional variations include the addition of various herbs and/or paprika. It is generally eaten with bread as part of a main meal. It has good keeping qualities. The lactic acid content is 1.6–2.5%.

9.4 Butter

Butter is one of the most important manufactured dairy products in the United Kingdom and Western Europe. It is derived from the agglomeration of butterfat globules, with the release of watery buttermilk when cream is worked in the churning process. The resultant product is a water-in-oil emulsion containing not less than 80% butterfat. There are two main types of butter, sweet cream and cultured cream. Either type may be salted at up to 2%.

9.4.1 Sweet Cream Butter

The cream for sweet cream butter manufacture is standardised to around 40% fat, pasteurised, cooled to 5°C and held for several hours to crystallise the fat. This crystallisation is an important step in order to achieve efficiency in the subsequent churning (mixing) stage. The cooling period is known as the 'ageing period'. After ageing, the cream is warmed to 10°C for churning where the fat globules are broken and the fat aggregates into grains. The fat binds together and water is released in the form of buttermilk. The fatty mass is then worked to achieve a good smooth texture and salted if required. In the United Kingdom, the remaining water level must not exceed 16% [14].

Buttermilk has a similar composition to skimmed milk, but possesses good emulsifying properties due to its high phospholipid content. It is commonly used in dairy analogues, baked goods and ice-cream.

9.4.2 Cultured Cream Butter (Lactic Butter)

Lactic butter is made from cream which has been 'ripened' by lactic acid bacteria, yielding lactic acid and diacetyl, which are largely responsible for its characteristic taste. Most lactic butter is unsalted, but it may be salted at up to 0.5%. Lactic butter is very popular in France, Denmark and other European countries, but in the United Kingdom, it repesents only a small proportion of the butter market.

After pasteurisation, the cream is innoculated with a starter culture, which may include a combination of *Streptococcus lactis*, *S. cremoris*, *S. lactis* subsp. *diacetylactis* and *Leuconostoc cremoris*. It is ripened for a few hours, then the cream is cooled and held at 7°C for several hours before being churned. The acidity of the cream gives it different manufacturing qualities to sweet cream, which manifest themselves in faster churning and higher yields. However, lactic butter is more susceptible to oxidative rancidity than sweet cream butter, and the buttermilk produced is acid and less easily utilised. These disadvantages have been largely overcome by modern manufacturing processes in which a culture concentrate is injected at the working stage of butter manufacture, after the separation of the buttermilk.

9.4.3 Ghee

Ghee, commercial clarified butter, is made by heating butter to remove the water. The heating process causes Maillard-type reactions between fat and non-fat components of the butter, and this generates characteristically different flavour profiles from butter itself.

9.4.4 The Flavour of Butter

The flavour of any food is a combination of aroma, taste and mouthfeel. With an emulsion like butter, some flavour components reside mainly in the fat phase; others in the aqueous phase. The pH of the aqueous phase also affects this partition between the phases. This is obviously an important factor in cultured cream butter, where lactic acid is present in quite high concentration in the aqueous phase. Perception of individual flavour components also varies in different media. Generally, flavour compounds tend to taste stronger in aqueous media than in fatty media.

The flavour of butter is the result of a complex combination of the mouthfeel properties of the butterfat, with varying release properties of each of the flavour components according to their partition between the phases. Much work has been done on flavour release in emulsions by McNulty [15]. Threshold values of individual flavour components, and synergistic effects between them are very important for flavour release in emulsions [16].

Many compounds have been identified in butters of various types. Useful qualitative and quantitative listings are given by CIVO/TNO [11]. It is interesting to note that of the 233 different compounds listed by TNO, 61 are carbonyls, 21 are fatty acids and 24 are lactones. These three groups represent by far the most important flavour contributors to butter character. Lactones in particular seem to be of great importance in butter flavour, and much work has gone into their identification [17]. In general, a review of the literature would suggest that the flavour of cultured cream butter is well understood, whilst that of sweet cream butter has been studied in less depth. However, this is debatable and the level of understanding of the development of flavour in the natural foodstuff is not necessarily directly related to the quality of the flavourings that the flavourist can develop. Cost effectiveness, raw material availability (with the right status) and an appreciation of the processing to which the flavouring is to be subjected are all also of vital importance in the development of a new flavouring.

Fatty acids are key materials in butter flavour, both in their own right and as precursors to other important compounds. Most important from a flavourist's point of view are the volatile acids (acetic, propionic, butyric, hexanoic and octanoic) and of course lactic acid in cultured cream butter.

The other important class of compounds which contribute to butter flavour is carbonyls; aldehydes, ketones and diketones. The best known of these is probably diacetyl, which with acetaldehyde, is produced during the cream ripening stage of cultured butter manufacture. Both compounds also play a smaller part in the flavour of sweet cream butter. A number of other short chain aldehydes, particularly unsaturated compounds, are also important in butter flavour.

Many alcohols, largely homologous with the ranges of acids and aldehydes present, have also been identified in butter. Their presence in equilibrium with acids give rise to low levels of esters which are often responsible for fruity off-flavours. Low levels of very low odour threshold compounds such as indole and dimethyl sulphide [18] also have a key role in some types of butter.

Besides the basic distinction between sweet cream and cultured cream butter, there are many other geographical differences. In cultured cream butter, the exact strain(s) of bacteria in the starter culture greatly affect the development of flavour, so that Danish lactic butter is quite different from German sour cream butter. There are similar differences between sweet cream butters, for example Dutch butter is generally much creamier than English butter, which normally has a richer, more aged character.

When butter is heated its flavour becomes stronger. Therefore in virtually all processed food products in which it is an ingredient, this effect is noticed. Aroma compound development is encouraged in both the fat and aqueous phases. The levels of volatile fatty acids increase in the fat phase, with a consequent increase in lactones (formed by cyclisation of hydroxyacids). In the water phase, Maillard-type reactions occur, leading to typical products such as furfural and other conjugated furans.

Thus, the flavour of butter is complicated, and cannot be reproduced with just a couple of key compounds. For example, diacetyl is important in all butters, but has been rather over used; customers quite often ask for a 'non-diacetyl type' butter flavouring!

9.4.5 The Uses of Butter Flavourings

Butter is used as an ingredient in many foods to provide both flavour and mouthfeel. Butter flavourings can be used in any application where butter might be used, but cannot either for technical or economic reasons. Some of the key applications are sauces, dressings, desserts, ice-cream, baked goods, margarines and sugar confectionery. The diversity of applications requires a diversity of butter flavour types. Not only does the required butter character change with geography and processing (see above), but the required physical form, legal status, solubility and heat and freeze-thaw stability also vary from project to project. The range of common application dosages varies enormously, from as little as 0.001% (10 ppm) up to 1% or more.

Flavourings for margarine tend to be very low dosage (0.001–0.02%). They may need to be oil- or water-soluble (especially for low-fat spreads), and tend to be liquids, pastes or emulsions.

Baked goods must rely on less volatile flavouring materials. Common dosage for nature identical or artificial flavourings is around 0.2%, but it may be up to 1% or more for naturals. The physical form will depend upon the processing involved and the customer's convenience of handling.

High-acid dressings are often improved by the addition of a creamy butter flavour to offset the acidity without affecting the keeping quality of the product.

9.4.6 Margarine and Low-Fat Spreads

Margarine is an analogue of butter. It consists of an emulsion of animal and/or vegetable oils, with an aqueous phase which may contain dairy and non-dairy solids, and sometimes salt. It has no desirable inherent flavour, and thus must be flavoured with a suitable butter flavouring. In recent years there has been much development in this product area, particularly in low-fat spreads, dairy blends and high-polyunsaturated products. This has put demands on the flavour industry to supply new flavourings to suit these innovative products. For example, flavourings for low-fat spreads are often expected to mask the unpleasant gelatine character of the aqueous base. Also, customers often wish to apply flavourings to both the aqueous and fat phases, although the authors believe that whichever phase the flavour is applied to, the individual flavouring compounds will, in an equilibrium distribution, partition between the phases according to their relative solubilities.

9.4.7 *The Development of Butter Flavourings*

The preceding discussions determine that it is essential to know the market into which the product to be flavoured will be launched. In addition to geographical variations, in some countries there are restrictions on the use of butter flavourings, particularly in respect of labelling. Also, as with all well-established flavour types, there are generally accepted, but unrealistic butter flavourings, which have gained credibility through many decades of use.

Choice of solvent system or powder carrier restricts the use of certain ingredients: lactic acid will not dissolve in vegetable oils because of its water content. High levels of decanoic acid will not dissolve in propylene glycol. There are many other such examples of technical constraint.

A wide range of flavouring substances are available for the development of butter flavourings. They range from the now less commonly used artificials, through a wide range of nature identical synthetics, to an increasing number of natural substances and natural preparations. When artificials are acceptable, the two most commonly used are butyl butyryl-lactate (for tenacious creamy body) and ethyl vanillin (for sweetness).

Amongst the wide range of nature identical and natural substances the key types are:

fatty acids	for bite, mouthfeel, roundness
lactic acid	for acidity
carbonyls	for creaminess, freshness, caramel notes
lactones	for creaminess, nutty characters
sulphur compounds	for heated butter character
alcohols and esters	for fruity notes

Lipolysed butterfat, obtained either by enzymic or acid hydrolysis of butterfat is a useful building block for the development of natural butter flavourings. A huge range of other natural flavouring preparations find use in the development of natural butter flavourings.

As is the case for most dairy flavourings, the enormous number of ingredients which could find use in butter flavourings precludes the construction of a comprehensive list. Almost any aromatic raw material could be useful at some stage in the development of a flavouring to meet a particular market/processing need defined by the customer. The flavourist certainly should never be bound by the limits of those compounds known to exist in real butter.

9.5 Cheese

The term cheese can be applied to dairy foods varying from a mild white acidic paste to an intensely aromatic brown waxy mass. Cheeses are generally nutritionally rich (in calcium, protein, fat and vitamins) and organoleptically rich (in intensity and variety). Cheese is eaten cold or hot; and in savoury and sweet dishes. Yet to some it is nothing more than rotten milk!

Technically, cheese is the product of coagulation of casein and removal of water from milk. The milk source may be whole, semi- or fully skimmed milk, cream or buttermilk; or any combination of these. The removal of water is what makes cheese significantly different from other coagulated milk products such as yogurt.

The history of cheese can be traced back a very long way. Archaeological signs of domestication of dairy animals date back to 6500 BC. By 3500 BC, artwork shows scenes of milking animals and the keeping of milk, and by 3000 BC definite proof of cheese-making is evident in Egypt, Greece and Italy [19]. A rich history of the progression of cheese is well documented to the present day, when almost every country in the world makes some kind of cheese [20].

It is in the area of cheese that the variety of flavour characters that are derived from milk shows its true immensity. This variation itself determines that we can only touch the surface of the subject in this chapter. Fortunately, reasonable classifications by basic flavour type are possible, and this will enable us to give an overview which should help the flavourist faced with the job of reproducing the character of a cheese of which he has never heard, let alone tasted!

9.5.1 The Manufacture of Cheese

The methods of cheese manufacture depend greatly on the type of cheese being produced. However, certain basic steps are fairly common. The three fundamentals are coagulation of milk protein, separation of solids, and working on the solids to modify flavour and texture. These basic steps are represented in Figure 9.4 by solid arrows. The steps indicated by broken arrows are undertaken for only some types of cheese, and their sequence in the production process varies depending on the type of cheese. Innoculation with bacteria, yeasts and moulds is done at many different stages in the process, and this greatly affects flavour and texture development. In industrial scale manufacture, such innoculation is scientifically controlled, whilst in the traditional farmhouse it may occur directly from the environment: in fact many cheeses owe their existence to such 'accidental contamination'.

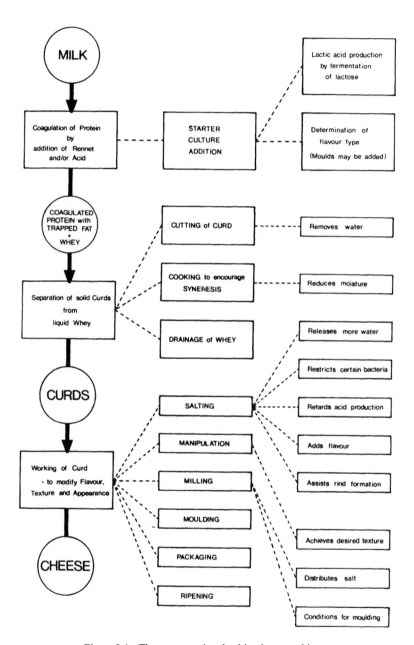

Figure 9.4 The processes involved in cheese-making.

Table 9.1   Classification of cheeses by ripening, texture and flavour

Ripening	Texture	Examples	Source[a]	Flavour type[b]
Unripened	Soft	Cottage	Skimmed	Diacetyl/lactic
		Ricotta	Whey	Diacetyl/lactic
		Mascarpone	Cream	Creamy (heated)
		Neufchatel	Cream	Creamy/lactic
Unripened	Plastic	Mozzarella	Buffalo	Fresh/lactic
Ripened	Soft	Reblochon	Whole	Buttery
		Chabicou	Goats'	Goaty
		Feta	Mixed	Salty/ketones
Ripened	Semi-hard	Caerphilly	Whole	Diacetyl/lactic
		Gouda	Whole	VFA/waxy
		Saint-Paulin	Whole	Amino acids/bitter
Ripened	Hard	Cheddar	Whole	VFA/S-cpds:complex
		Lancashire	Whole	Diacetyl/lactic
		Cheshire	Whole	Diacetyl/lactic
		Parmesan	Whole	Strong VFA:complex
		Romano	Ewes'	Strong VFA:complex
		Emmental	Whole	VFA/nutty:complex
Surface mould	Soft	Camembert	Whole	VFA/NH_3:complex
		Brie	Whole	VFA/NH_3:complex
Surface smear	Soft	Limburger	Whole	VFA/S-cpds:complex
		Munster	Whole	VFA/S-cpds:complex
Internal mould	Semi-hard	Stilton	Whole	VFA/ketones
(blue)		Roquefort	Ewes'	VFA/ketones
	Soft	Gorgonzola	Whole	VFA/ketones

[a] Source is cows' milk unless otherwise stated.
[b] Flavour type—broad indications only: VFA, volatile fatty acids; S-cpds, sulphur compounds; NH_3, amino compounds.

9.5.2   Classification of Cheese Types

At this point, we intend to take the contentious step of offering a broad classification of cheeses by texture, ripening and flavour type. Such generalisation will be certain to offend some readers, so we will again emphasise our concentration upon the flavour character of the individual cheeses (Table 9.1).

9.5.3   The Development of Flavour in Cheese

Complex biochemical and chemical processes, combining the breakdown of curd by enzyme catalysed reactions, with further chemical interactions, give rise to the typical flavours, textures and appearances associated with particular types of cheese.

Lipolysis, proteolysis and fermentation each play significant roles in the development of cheese flavour. It should not be assumed that only volatile products are important in cheese flavour; the flavour of most cheeses is a complex blend of taste and aroma components, which results from the reaction, intermediate production, and further reaction of many components.

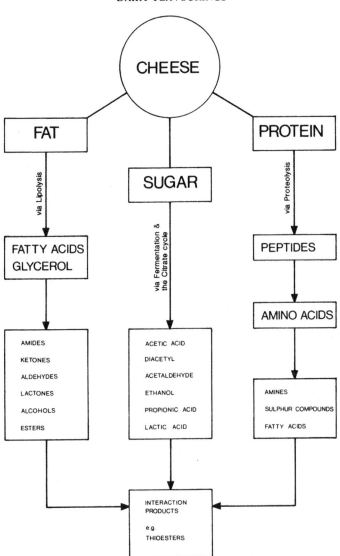

Figure 9.5 Flavour development in cheese.

The perception of flavour is a result not only of the relative proportions of these components, but also of synergistic effects between them. In Figure 9.5 we have attempted to give a simplified overview of some of the chemical pathways in cheese flavour development. Those with a deeper interest are directed to the literature [21–23].

Fatty acids in cheese are derived via a direct route from milk, and it is in this area that the differences between species are particularly evident. Comparison of the saturated fatty acid contents of cows', goats' and ewes' milks gives an interesting insight into possible explanation of the differences between their respective types of cheese [24]. Reference to Figure 9.5 shows the importance of fatty acids as precursors of many of the compounds responsible for characteristic flavour in cheese.

9.5.4   Review of a Range of Key Cheese Types

As mentioned already, in this chapter it will be impossible to cover all, or even a large proportion of the cheeses available in the world. However, a number of the most important types in the Western World are now briefly discussed. They are presented in alphabetical order by name for easy reference.

9.5.4.1   Bel Paese.   Bel Paese is a soft uncooked Italian cows' milk cheese with an 80-year history [25]. Its aroma is slightly sulphurous, acidic and reminiscent of sweaty feet! The taste is acidic and creamy, with a pleasant slightly bitter aftertaste. The better quality the cheese, the less bitterness is evident.

9.5.4.2   Brie and Camembert.   These are French soft cheeses made using mesophilic starter cultures. The curd is uncooked, leaving its moisture content high and the characteristic flavour derives from the white mould which grows on its surface. Spores of this mould (*Penicillium candidum, P. caseicolum* and/or *P. camemberti*) are seeded onto the surface of the cheeses after dry salting, and flavour development occurs by proteolysis and lipolysis from the surface inwards. In addition, naturally occurring environmental yeasts contaminate the surface and add to flavour development.

Brie has a creamy, slightly acidic, nutty flavour with characteristic musty, mushroom-like ammoniacal overtones. Camembert is matured longer than Brie, and develops a secondary surface flora of *Brevibacterium linens*, which gives rise to other strongly aromatic compounds including methanethiol. An important volatile contributor to the mushroom-like notes of both these cheeses is 1-octen-3-ol [26].

9.5.4.3   Blue-veined cheeses.   'Blue-veined' describes the appearance of those cheeses, the body of which is veined with growths of blue-green *Penicillium* moulds. In the manufacturing process, the curds are innoculated with spores, then when air is admitted to the fresh cheeses by piercing, the spores germinate and spread throughout the cheese. The mould assists in flavour development, being strongly proteolytic and lipolytic, and adding the effect of the starter culture.

The characteristic flavour of blue-veined cheeses is a combination of short-chain fatty acids and methyl ketones. The action of the mould releases fatty acids, which are further degraded to methyl ketones, which are largely responsible for the fruity character of such cheeses. Extreme proteolysis is also common in such mould-ripened cheese, and this gives characteristic bitterness [27].

9.5.4.4 *Cabrales (picon)*. Cabrales is a hard Spanish blue-veined cheese made with cows' milk. The cheese is matured for up to 6 months in limestone caves in the Asturias, where it develops a sharp and distinctive blue-veined flavour.

9.5.4.5 *Caerphilly*. Caerphilly is a crumbly, semi-hard white cheese origi-nating in Wales. Made from whole cows' milk, it has an acidic flavour, with slight reminiscences of cowsheds.

9.5.4.6 *Chaumes*. Chaumes is French, unpasteurised cows' milk cheese, which is soft-textured, golden creamy in colour, with a brown rind. The strong aroma is of old socks, but the taste is milder; light and creamy.

9.5.4.7 *Cheddar*. Cheddar is probably manufactured in more countries throughout the world than any other type of cheese. Consequently, there is much literature on the flavour of Cheddar cheese; but there are few really good Cheddar-type cheese flavourings available! Cheddar has a very complex flavour profile which is not easily summarised in terms of individual aromatic compounds. In application, its flavour has quite different characters: cooked Cheddar develops a significantly changed flavour during the heating process. The common availability of Cheddar at various stages of maturation also adds to the variation in flavour, and the consequent difficulty in characterisation.

The literature indicates the importance to Cheddar of fatty acids, methyl ketones, amines and sulphur compounds, especially methanethiol. Its flavour has a creamy, acidic, sulphurous character. Some of the important volatiles may be hydrogen sulphide, methanethiol, diacetyl, butanone, pentan-2-one, dimethyl disulphide, acetone and ethanol [28]. Some esters may be present, but high levels are detrimental to the flavour. None of these compounds is, in itself, characteristic of the complex flavour of Cheddar, and this is reflected by the difficulty in developing realistic Cheddar flavourings.

9.5.4.8 *Cheshire*. White and red Cheshire, the latter simply coloured with annatto, are hard cows' milk cheeses, which have a mild acidic, almost buttery, slightly salty flavour. Blue-veined Cheshire is stronger in flavour with an additional bitter acidic tang.

9.5.4.9 *Cottage cheese*. Cottage cheese is a soft white, lumpy, watery cheese, which is sold and eaten in the fresh, unripened state. It is made from skimmed milk which is acidified with a mixed starter culture of lactic acid bacteria.

Rennet may be added to provide the clotting agent chymosin, but it is not fundamentally responsible for flavour in unripened cheese. The flavour is derived from fermentation products.

The acidity of the cheese is provided by lactic acid; diacetyl is produced by *Streptococcus diacetylactis* from milk citrate. Acetaldehyde is also a product of fermentation, but in a well-flavoured cottage cheese the level of acetaldehyde should be well below that of diacetyl; the reverse situation to that pertaining in cows' milk yogurt.

9.5.4.10   *Cream cheese.*   Cream cheese is made from a blend of milk and cream. It has a smooth and buttery texture, with a creamy taste. It is often sold with added flavour and textural ingredients, such as pineapple and chives.

9.5.4.11   *Doux de montagne.*   This French cheese, originating from the Pyrenees region, is semi-hard, mild and creamy. It has slightly lactic, creamy aroma with a hint of sourness. The taste is creamy, buttery, lactonic, slightly caramel with a hint of bitterness.

9.5.4.12   *Danablu.*   Danablu is the Danish trade name for their cows' milk equivalent of Roquefort. This whole cows' milk, blue-veined semi-hard cheese has a strong, savoury, typically ketonic/acid blue cheese character.

9.5.4.13   *Double Gloucester.*   This pale, orange-coloured, smooth-textured hard cheese has a milder flavour than many Cheddars, but is broadly similar to Cheddar in both production and flavour.

9.5.4.14   *Feta.*   Feta is a cheese of Greek origin, which is sold in blocks stored in brine. The taste of this white cheese is sharp and salty. Originally made from ewes' milk, Feta is now made in many countries from milk of many animals. The flavour varies in profile according to the milk used [29].

9.5.4.15   *Fromage frais.*   This term is applied to a group of very soft moist cheeses made from pasteurised cows' milk. The taste is mild, fresh and acidic, and these cheeses are often eaten with fruit, and even sold fruit-flavoured. Quarg, which is common across northern Europe, is a member of this group. These cheeses can be made with whole or skimmed milk, resulting in different fat content with associated differences in flavour and mouthfeel.

9.5.4.16   *Gjetost.*   Gjetost is made from goats' milk whey in Scandinavia. There are hard and soft varieties, but neither is commonly available in the United Kingdom. Strictly, it is not a true cheese, as the whey is cooked and evaporated prior to cheese production. Gjetost has a caramellised note and resembles sweetened condensed milk or fudge.

9.5.4.17 *Gorgonzola.* Gorgonzola is an Italian soft blue-veined cheese, made with uncooked whole cows' milk. The bluish-green internal veining is the result of growth of *Penicillium glaucum*, a different strain from that used in Roquefort and most other blue cheeses. Gorgonzola has a very salty, tangy, almost 'mouldy' blue flavour, with a slightly dirty aroma. It is an excellent cheese when at its best.

9.5.4.18 *Gouda and Edam.* These are Dutch semi-hard cows' milk cheeses. Both are available in relatively mildly flavoured young form, or more mature. In both cases, the rindy, 'waxy' notes increase with age. Gouda has a fuller flavour due to its higher fat content, whilst Edam has a salty note derived from its brine treatment.

9.5.4.19 *Havarti.* This is a Danish semi-hard cows' milk cheese which has small eyes, with a fairly mild flavour reminiscent of mild Dutch cheeses.

9.5.4.20 *Italian Romano, Provolone and Parmesan.* The efficiency of degradation of triglycerides in cheese manufacture greatly affects flavour development. In these three Italian cheeses, distinctive flavour is generated by the deliberate addition of animal lipases, which generate low molecular weight free fatty acids. Some other important volatiles which also contribute to the characteristic Italian hard cheese flavour are ethyl hexanoate, 2-heptanone and pentan-2-ol [30].

Provolone is a plastic spun curd cheese. Unless well aged it is mild, but tangy cheese with a slight phenolic hint.

Pecorino Romano is a cooked curd ewes' milk hard cheese. It is full flavoured and sharp, with typical ewes' milk aftertaste. Romano is the name used in the United States for the domestic equivalent hard cows' milk cheese, made by the Pecorino Romano technique.

9.5.4.21 *Jarlsberg.* A softer, sweeter variation of Emmental, which is made in Norway.

9.5.4.22 *Limburger.* Limburger is a semi-soft cheese originating from Belgium. Its flavour is influenced greatly by the growth of microorganisms on its surface during ripening. The principal microorganisms are yeasts, *Micrococci* and *Brevibacterium linens*. Major flavour contributors are phenol (which arises from microbial decomposition of tyrosine), dimethyl disulphide and indole [31]. A very smelly cheese!

9.5.4.23 *Lancashire.* A pale creamy soft cheese which is moist and crumbly with a delicate salty, creamy flavour. A good toasting cheese.

9.5.4.24 *Leicester.* A hard cheese with a grainy texture, made by a process similar to that of Cheddar. It has a mellow, sweet and nutty Cheddar-type flavour.

9.5.4.25 *Manchego.* A Spanish ewes' milk hard cheese, the flavour of which is distinctive and sharp, and varies in strength with age.

9.5.4.26 *Mascarpone.* An Italian soft cream cheese which looks like whipped double cream. The taste is like slightly sweetened double cream, with a hint of cooked milk character.

9.5.4.27 *Mozzarella.* Mozzarella is the traditional plastic curd cheese used on Italian pizza. Traditionally made from buffalo milk, it has a very mild, slightly acidic buttery flavour.

9.5.4.28 *Munster.* Munster is a French, cows' milk soft cheese. Its flavour is influenced by the growth of microflora on its orange/yellow rind. It has a delicate salty flavour when young, which develops into a full tangy flavour after maturation.

9.5.4.29 *Neufchatel.* Neufchatel is a soft cheese, which is eaten young when its flavour is like tangy cream cheese.

9.5.4.30 *Pont l'Eveque.* This is a French, cows' milk soft cheese, which has the typical strong smell associated with washed rind cheeses. It has a rich creamy taste, characterised by methyl thioesters, phenol, creosol and acetophenone [32].

9.5.4.31 *Port-Salut.* Virtually indistinguishable from Saint Paulin.

9.5.4.32 *Quarg.* Although historically linked with Germany, Quarg is already widespread across most of Europe, and is gaining popularity in the United Kingdom. It is a soft acid-curd cheese with a bland but acidic flavour.

9.5.4.33 *Queso blanco.* This is a white cheese usually eaten fresh in Latin America. It can also be vacuum packed for a one year shelf-life. Queso blanco can be made in a variety of ways; starter and rennet, or by addition of acids with heating.
There are several varieties of Queso blanco, determined by variations in salting or addition of different microorganisms. High temperature storage results in butyric notes from fermentation of lactose. It can also be flavoured at various stages in manufacture by addition of flavouring ingredients such as fatty acids and enzyme-modified cheese [33].

9.5.4.34 *Reblochon.* Reblochon is a French, cows' milk soft cheese with a particularly creamy, buttery flavour.

9.5.4.35 *Ricotta.* This is not strictly a cheese at all. Whey is re-cooked to precipitate the proteins which form the basis of the cheese. Ricotta has some similarity to cottage cheese, but is slightly fuller in both texture and flavour.

9.5.4.36 *Ridder.* Ridder is a semi-hard cows' milk cheese made in Scandinavia. It smells of sweaty feet in the Gouda/Edam sense; with a creamy diacetyl, New Zealand butter character. In taste it is rather like a young Gouda, with a hint of caramel or cane syrup.

9.5.4.37 *Roquefort.* The famous ewes' milk blue-veined cheese matured in the limestone caves of Cambalou at Roquefort. The curd is innoculated with spores of *Penicillium roqueforti* which grows throughout the cheese to form bluish-green veins. Roquefort has a distinctive ketonic blue cheese taste, with buttery undertones, and a characteristic ewes' milk tang. The flavour is very powerful.

9.5.4.38 *Saint Paulin.* A French cheese which is innoculated with a lactic acid bacteria culture, which increases the acidity and leads to its distinctive character. It smells and tastes of hydrolysed casein.

9.5.4.39 *Samso.* A Danish cheese with a mild aromatic, sweet flavour.

9.5.4.40 *Smoked cheese.* When smoke flavouring is introduced into the cheese during manufacture the resulting product is often referred to as smoked cheese. There are many such examples of smoked cheese from around the world.

9.5.4.41 *Stilton.* The most famous English blue-veined cheese, Stilton has a sharp acidic, slightly bitter character, with a pungent, almost cowshed-like aroma. *Penicillium roqueforti* is used in its manufacture.

9.5.4.42 *Swiss Emmental and Gruyere.* Emmental and Gruyere are the best known Swiss cheeses in the United Kingdom. They are hard ripened cheeses, similar to, but differing significantly from Cheddar. Bacteria involved during ripening include propionic acid bacteria, which generate propionic acid. This acid plays an important role in the aroma of these cheeses, along with acetic and lactic acids. A further product of propionic acid bacteria fermentation is carbon dioxide, which is responsible for the large holes in these cheeses. A good Emmental cheese has a characteristic nutty flavour. Several components are said to be important; acetic acid, propionic acid, butyric acid, proline and diacetyl [34].

Gruyere generally has a 'dirtier' flavour than Emmental, especially near the surface. The mountain varieties of Gruyere have flavour derived from surface flora growth such as lactate-utilising yeasts. These lift the pH at the surface, so allowing other microorganisms, for example *Brevibacterium linens*, an orange pigmented coryneform, to grow. The mechanisms involved in the development of flavour are very complex. The putrid character is partly attributed to methanethiol and thioesters of acetic acid and propionic acid. The flavour does vary considerably through the cheese from rind to centre, as demonstrated by the relative abundance of aromatic compounds analysed in different regions of the cheese [35, 36]. This variation makes it particularly important for the flavourist to establish exactly what type of Gruyere character a customer is requesting; the description 'Gruyere' is insufficient.

9.5.4.43 *Taleggio.* This is a surface-ripened, uncooked curd soft Italian cheese. It has a delicate mildly acidic, buttery flavour.

9.5.4.44 *Tilsit.* This is a semi-hard cows' milk cheese, originating from Germany. Many aromatic compounds have been identified in Tilsit. Key flavour contributors include the branched chain fatty acids, isobutyric, isovaleric and isohexanoic [37].

9.5.4.45 *Wensleydale.* A white crumbly cheese with a nutty, creamy taste; best eaten fairly young.

9.5.5 Related Products

A number of other related products should be mentioned at this stage. In processed cheese, typically, hard and/or semi-hard cheeses such as Cheddar and Gouda are blended with skimmed milk powder, butter and other ingredients, then heated by direct steam injection with emulsifying salts to produce a homogeneous cheese-like mass. The proportions can be varied to achieve the desired consistency for spreads, slices, etc. If the temperature/time treatment is sufficient, pasteurisation of the product can be achieved, giving valuable extension to shelf-life, even under ambient conditions. Re-melting of processed cheeses for culinary uses is also achieved easily and without fat separation, because of the emulsifying salts.

The expense of milk fat in some parts of the world has led to the development of cheese analogues, in which a cheese-like product is made using casein (usually derived by rennet coagulation) and cheaper vegetable fats. Normal cheese flavour development does not occur in such products, because the key fatty acids are not present. Therefore cheese flavourings are usually added to achieve a desirable product. Such materials find use in food processing, especially for pizzas, where there are cost and handling problems in using real cheese (e.g. Mozzarella).

The development of spray drying has enabled the production of dehydrated cheese powders and cheese powder compounds (blends of cheese, vegetable fat and other ingredients including flavourings). Such products are of great value to the food industry since they eliminate the refrigeration and handling problems associated with 'wet' cheese.

9.5.6   Applications of Cheese Flavourings

The most obvious applications for cheese flavourings are sauces, dips, baked goods, fillings and snack foods; but the range of products in which cheese flavourings find use continues to surprise even those working in the industry! Cheese flavourings are not permitted in processed cheeses in many countries, but in others this is an important application area; whilst cheese analogues could not exist without flavourings. There is also some application in cheese powder compounds.

In the United Kingdom, the most often requested cheese flavouring type is not surprisingly Cheddar, probably followed by the blue-veined types. However, with the integration of the United Kingdom more and more into the European Community, the range of other cheese types requested is growing rapidly.

9.5.7   The Development of Cheese Flavourings

There is much in the literature about the flavour of cheese, but with the odd exception [38], little of value about the development of cheese flavourings. This is almost certainly because of the commercial sensitivity of flavouring formulations. This chapter will be no exception to this general principle, but we shall try to highlight the key points for consideration by the flavourist embarking on the creation of a cheese flavouring.

As with any flavour area it is useful for the flavourist to determine a set of essential characteristics and then decide on the relative abundance of each. Figure 9.6 sets out one possible way of profiling a cheese flavour. This assumes that the individual flavourist chooses a compound or set of compounds for each flavour element; e.g. 'sicky acidic' = butyric acid.

Thus, in this example, Danablu cheese has a predominantly soapy and acid character, with emphasis on the sharp and sicky acidic notes. Soapiness (and some fruity notes) can be achieved with methyl ketones; acidic character with fatty acids; and fruitiness with esters. The saltiness should not be forgotten, and in order to achieve a realistic effect in the end-product, salt will either have to be incorporated in the flavouring or added to the end-product.

Important flavouring materials for use in cheese flavourings are cited in the literature; fatty acids, ketones (especially methyl ketones), aldehydes, lactones, alcohols, esters, amines, sulphur compounds. These are the volatile components. Also vitally important in cheese flavour are the lipids, lipoproteins, peptides and amino acids. By having some knowledge of cheese manufacture and its relationship to the cheese type produced, it is even possible to 'guess' a flavouring formulation without sampling the cheese itself. This is clearly an

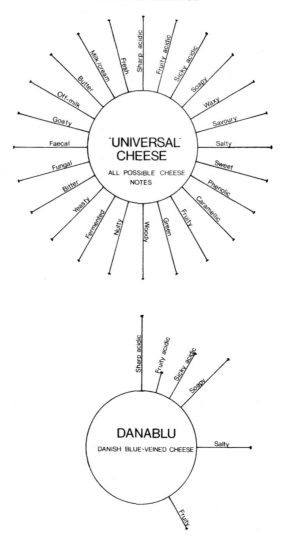

Figure 9.6 Flavour profile diagrams for a 'Universal' cheese and for Danablu.

important factor for the flavourist who is asked to service the cheese flavouring needs of parts of the world, whose local cheeses are not familiar.

The choice of ingredients open to the flavourist is very wide, since many of the compounds identified in cheeses are now available in nature identical form; and even, for a price, natural form. For example, natural butyric acid is currently about 25 times the price of the nature identical material.

Despite this wide range of available single chemicals, the flavourist is unlikely to be able to formulate realistic and cost-effective cheese flavourings with these alone. A wide range of natural flavouring preparations can also be used in cheese flavourings. These range from foods themselves, such as cheese powders, through many extracts, to microbiological preparations such as enzyme-modified cheese (conventionally prepared cheese to which lipases have been added to accelerate ripening), and other enzymic preparations of dairy and other materials.

9.6 Manufacturing Considerations

When developing any new flavouring, the flavourist has two main customers to consider. The product development technologist in the customer company is the first priority. However, the flavour house's own production department is also a vital customer of the flavourist's work. It is no good the flavourist developing a flavouring which totally fulfils the needs of the brief, in terms of character, dosage, price and effectiveness, if the production department are unable to scale the product up from the bench level at which it was developed! The flavourist must have a good appreciation of the needs, methods, equipment and capacity of the production department in order to provide them with a formulation which can be turned into a saleable product.

The developed flavour should be organoleptically repeatable when scaled-up, and as simple as possible to compound. The formulation should be as concise as possible, using only reliable and approved suppliers, and without redundant ingredients. The flavouring must be easy to produce repeatably within practical and sensible specifications, as well as meeting the total requirements of the customer company.

Particular problems to be considered in the production of dairy flavours include:

unpleasant smell
difficult and hazardous materials to handle, e.g. fatty acids, lactones, acetaldehyde
flammability, e.g. diacetyl
possible biological activity in dairy ingredients
many ingredients and solvents are oily, spills can lead to very slippery floors

Generally, these problems and hazards (to both the product and the operatives) can be overcome by strict adherence to good manufacturing practice [39] and normal health and safety rules.

9.7 Conclusion

Dairy flavourings is an interesting and potentially very profitable niche in the flavour market. It shares many of the opportunities and constraints of the rest of the flavour industry, but provides an interesting self-contained challenge to the flavourist. Understanding of dairy flavours is less well-developed than that of fruit flavours, for example, and the relative complexity of the flavours and the multiphase systems in which they develop, add to the intellectual challenge. On a business level the market place is less saturated with competitive flavourings than in many other flavour areas. The dairy area has much to interest flavourists who wish to broaden their experience.

10 Process flavourings

C.G. MAY

10.1 Introduction

10.1.1 *Description and Definition*

'Reaction' or 'Thermal process' flavouring is a comparatively recent term given to a food flavouring which is produced by heating together two or more 'precursors' under carefully controlled conditions. If desired, other flavouring materials may be added after the reaction to make a composite blended flavour. The precursors may have little flavour in themselves, this being developed or produced by a process analogous to that used for the cooking of everyday foods. Since their introduction to the food industry, they have been widely known as 'reaction' flavourings. More recently, various legislative bodies have adopted the name 'thermal process' or 'process' flavouring.

The 'reaction' flavouring differs from other more traditional types of flavouring, which may be either extracted from natural sources or blended mixtures of chemicals in which heat is usually kept to a minimum and is not part of the flavour development. The chemical reactions involved in producing reaction flavourings are exceedingly complex, and will be discussed briefly later.

The heating together of isolated precursors under laboratory or factory conditions to give concentrated reaction meat flavourings has been known to the flavour and food industry for only about 35 years, and this class of flavour is proving difficult to define or understand. For impending Safety Legislation, more precise attempts at definition have been made. The International Organisation of the Flavour Industry (I.O.F.I.) has issued a Code of Practice, which includes a section 'I.O.F.I. Guidelines for the Production and Labelling of Process Flavourings' in which is the definition: 'A thermal process flavouring is a product prepared for its flavouring properties by heating food ingredients and/or ingredients which are permitted for use in foodstuffs or in process flavourings.' Details are given of the permitted raw materials and ingredients and the recommended reaction parameters. These guidelines are given in Appendix III. A Council Directive in the Journal of the European Communities [1] is similiar to this in most respects, and revised draft guidelines were issued for discussion at a Council of Europe meeting in

Strasbourg, 24–28 April, 1989. These are under review, and a further revised directive is expected shortly.

In this chapter, the term 'reaction flavouring' refers to the 'complete flavouring', and consists of the 'basic reaction flavouring' (the flavour product obtained by thermal processing) plus other flavouring materials which may be added after the reaction.

Reaction flavourings originated as a result of research into beef flavour, as described briefly later. They have been developed to include other meat and savoury flavourings. This chapter is concerned solely with meat reaction flavourings.

10.1.2 History and Necessity

Until the beginning of the 1939 War, the diet of most people in the United Kingdom had been fairly stable and most basic foods were readily available. The flavour industry concentrated on producing confectionery, fruit and other non-savoury flavours, although various yeast derivatives were available for supplying 'meaty' tastes. Compared with today, few new food products were introduced. The war years gave an impetus to attempts to produce processed and substitute foods, and imported novel types of foods were common in Britain. There was of course a shortage of protein and fat in the National diet.

The immediate post-war years gave little improvement in this deficiency. As a desperate move, the British Government embarked on a bold project, the 'Groundnuts Scheme' (officially organised by The Overseas Food Corporation) to produce groundnut oil. It was argued that this oil could reduce the fat shortage. In industry, attempts were made to utilise the protein of the groundnut. For this, it was feasible to imagine the production of a 'meat analogue', perhaps in the form of a meat loaf (forms of meat loaf were by then generally accepted by the British public). A process for preparing edible meat-like structures from texturised, spun protein had been described by Boyer [2]. To be acceptable, such a product would have to have a pleasant appearance and texture, and a good flavour, reminiscent of the meat it was intended to replace. There were no imitation beef (or any meat) flavours in existence at the time, and a poor flavour was given by beef extract, which was expensive and in short supply.

Research into the flavour of beef was started at the Unilever Laboratories in 1951, in Bedford, and there was research in progress in the United States and perhaps elsewhere. The Groundnuts Scheme was soon abandoned as a failure. Progress on meat analogues was slow but there was by then a rapidly growing interest and demand for convenience foods, and other novel forms of food. Imitation meat flavourings were needed urgently. This increasing desire and need for convenience and processed foods, with a corresponding decrease in consumption of the more stable foods has continued to the present day, and was stimulated by economic conditions such as inflation, the shortage of some raw materials and the development of new raw materials, and the growth of the supermarket retail outlet. Changing social patterns also have had a large influence on our food. Many people, for example, have not the time or inclination for lengthy preparation and cooking of more traditional meals.

The population of the world is of course increasing rapidly. Over 30 years ago many people in the Western World began voicing their concern of the approaching protein shortage and starvation in developing countries. As keeping beef cattle is a very inefficient way of obtaining protein (not much more than 10% of the plant protein consumed is converted into beef) and satisfactory land is often scarce, efforts were intensified to derive protein from known vegetable sources, (such as soybean, cottonseed, groundnut and sesame seed) and to explore completely new sources.

Work on new sources of protein appeared to have enormous potential and was pursued with much vigour. Considerable success was soon revealed in the field of microorganisms. Organisms such as bacteria, yeasts and fungi, which have a very rapid growth rate, were grown on cheap or waste materials and gave proteins without the need for conventional agricultural procedures. Being cultivated under controlled conditions, the processes were independent of the weather. The exploitation of these new proteins in the human diet has not been as effective as their originators possibly hoped.

Development of the meat analogue, however, has been considerable, and much work has been done to produce textured protein foods. The protein now used commercially is generally derived from soya, which contains all the essential amino acids. It may, however, be deficient in sulphur amino acids, but can be supplemented with cystine and methionine to satisfy nutritional requirements. The future of meat analogues from textured protein did look very promising, but the expected rapid growth over the last few years has not occurred, perhaps owing to beef factory farming production methods and the rise in general affluence in the West. The analogues continue to be used, however, in a steady market as products in their own right and as supplements and extenders in other foods. Their inherent flavour is bland or even unattractive, and the application of suitable imitation meat flavours can be most beneficial.

The wide range of convenience and process foods which are now on the market or are being developed are produced from mixtures of different materials, involving different processing techniques. Many need the addition of imitation meat flavours to make them acceptable, and in some cases these flavours have to be 'tailor-made' to allow for these processing conditions.

10.2 Research into Beef Flavour

Until 1954–1955, the flavour of 'meaty' food products was often enhanced by the addition of protein hydrolysates and autolysates, monosodium glutamate, spice blends and perhaps meat extract. These materials had no true meat flavour but were useful in 'stretching' any flavours already present in the food product. They were inadequate if there was no inherent meat flavour in the product in which reaction flavours were to prove most valuable.

The term 'meat flavour' embraces the flavour of most flesh foods, excluding sea foods, e.g. beef, chicken, mutton, pork and ham, and can be extended to include the less important veal, venison, etc. The most popular and desired flavour has always been beef, and investigation of beef flavour received most attention initially. Little work had been published, but Crocker [3] had postulated that the factors responsible for flavour were present more in the juice than the fibre of the meat and that most of the flavour could result from thermal degradation of proteins. All raw meats had weak, blood-like flavours, the characteristic flavours being developed on cooking, and on top of the basic meat flavours, there was an additional 'species' flavour.

Work on beef flavour at Bedford [4] used two different approaches: one was to isolate and identify the aroma arising from beef on cooking, and the other was to isolate and characterise the flavour precursors which gave rise to the basic beef flavour. A brief summary of this work follows.

Beef was cooked by stewing (where the meat temperature is about 100°C) and by roasting (where the meat temperature is typically 150–200°C on the outside and much lower on the inside). Many volatile compounds were identified and obtained from both methods of cooking. Significantly there was a high proportion of sulphur and carbonyl (including some furfural) compounds. It became obvious that in all ways of cooking, the strong basic cooked meaty flavour arose from precursors in raw beef. Therefore raw beef was extracted with cold water, and dialysed. The fraction having a molecular weight of less than about 200 produced a beef flavour on cooking. This dialysate was examined by paper chromatography both before and after cooking (the more sophisticated instruments and techniques of today were not available at that time). The uncooked extract contained about thirty amino acids and small peptides, one amino sugar, and three sugars. The amino acids fraction contained large amounts of cysteine, β-alanine, glutamic acid and tryptophan. The three sugars were identified as glucose, fructose and ribose, and the amino sugar as glucosamine. Quantitative paper chromatography of the cooked extract showed that cysteine and ribose had disappeared and some of the other compounds had been reduced. It was then shown that the ribose came by autolysis (from the ribonucleotides) after death of the animal, the amount of free ribose increasing in relation to time. It was deduced that ribose and cysteine and other amino acids were involved in the production of the basic flavour of beef. Heating these compounds together in aqueous solution did produce a quite recognisable beef flavour. This reaction was developed, using cysteine, amino acids or protein hydrolysates, and reducing sugars (or some derivatives of sugars) to give the first imitation meat reaction flavours [5]. Cooked beef flavour was assumed to be a combination of the products obtained by the reactions of the precursors and the evolved volatile compounds, with other compounds such as fats playing an important part. The flavour of other types of meats is based on this also.

10.2.1 Reactions of the Precursors

The water-soluble meat flavour precursors comprise many different classes of organic compounds, including peptides, amino acids, sugars and amino sugars, amines and nucleic acid derivatives. The reaction of amino acids and sugars (the Maillard reaction [6]), with its accompanying non-enzymatic browning and production of aromas, has of course been known and investigated for some time. The reactions involved are very complicated. It has been the subject of conferences [7], and many papers and reviews have been published [8]. It occurs in many foods, and is accelerated greatly by cooking. Many pathways have been postulated about the reactions which occur. Some of these are given in Figure 10.1, from a publication of Leatherhead Food R.A. [9].

This survey gives a general introduction to the chemistry of the amino acid-reducing sugar reactions, and describes the many classes of compounds which can be produced. Before this, Hodge [10] had attempted an explanation of the chemistry of the Maillard reaction. In model systems, he showed how the early reactions between the carbonyl group of a reducing sugar and the free group of an amino acid (or a protein) give a Schiff's base. This, via cyclisation and the Amadori rearrangement gives the 1-amino-1-deoxy-2-ketose. This is illustrated in Figure 10.2.

By different pathways, this ketose produces aldehydes, keto-aldehydes, dicarbonyls and reductones and flavour compounds such as 2-furaldelhydes. This results in dark brown material, and involves aldol condensation. Another reaction, the Strecker degradation, is vital to the production of a good flavour. In this reaction, amino acids lose one carbon atom (producing carbon dioxide) to become aldehydes. The carbonyl compound must contain a $-CO-CO-$ or a $-CO-(CH=CH)-CO$-grouping, and there must be an alpha amino group. Thus cysteine $HS-CH_2-CH(NH_2)COOH$ would become mercaptoacetaldehyde $HS-CH_2CHO$, and methionine $CH_3SCH_2CH_2CH(NH_2)COOH$ would become methional $CH_3SCH_2 \cdot CH_2CHO$.

This is obviously a shortened and much simplified view of part of a very complex reaction system, which continues to be the subject of intensive study. At a recent symposium [11] various aspects of the reaction were discussed, and the present state of understanding of the Maillard reaction has been given in a concise review by Ames [12]. The reaction is influenced markedly by experimental conditions such as temperature, time, pressure, pH and reactant proportions. These have to be manipulated carefully to obtain a desired flavour.

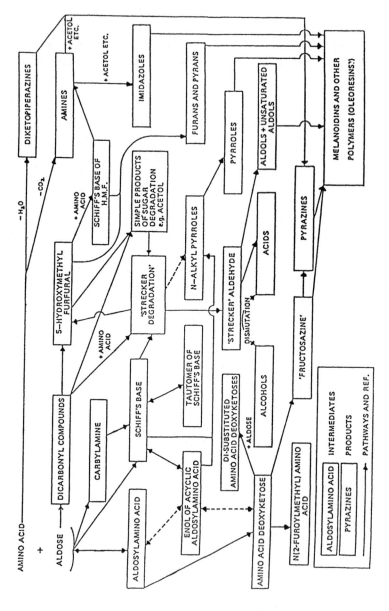

Figure 10.1 Possible reaction pathways of an amino acid-reducing sugar complex (from Stewart [9]). Reproduced by permission of the Director Leatherhead Food R.A.

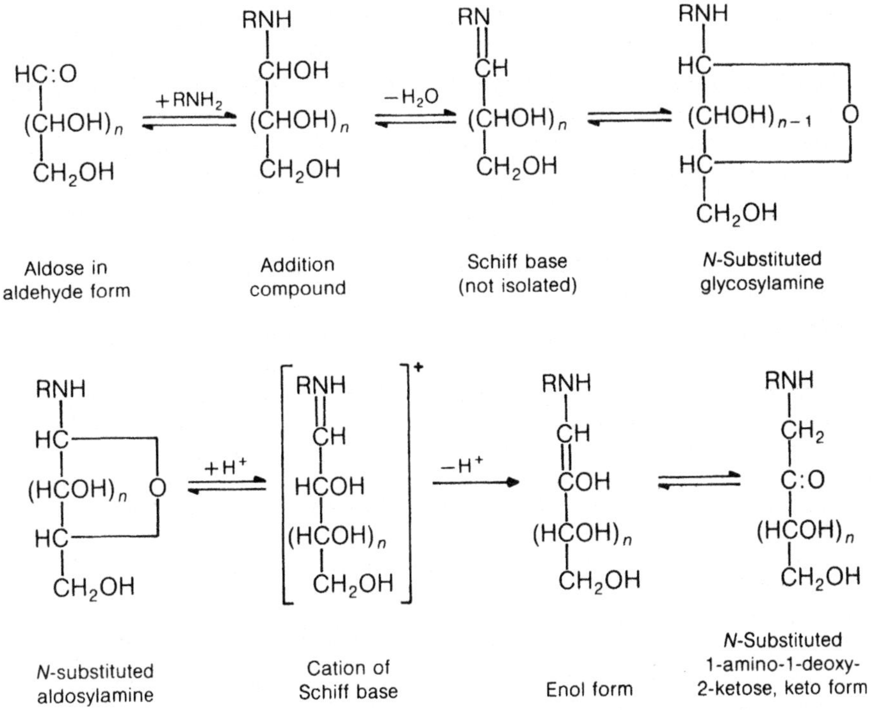

$$\underset{\substack{\text{Aldose in}\\\text{aldehyde form}}}{\begin{array}{c}\text{HC:O}\\|\\(\text{CHOH})_n\\|\\\text{CH}_2\text{OH}\end{array}}\ \underset{+\text{RNH}_2}{\rightleftharpoons}\ \underset{\substack{\text{Addition}\\\text{compound}}}{\begin{array}{c}\text{RNH}\\|\\\text{CHOH}\\|\\(\text{CHOH})_n\\|\\\text{CH}_2\text{OH}\end{array}}\ \underset{-\text{H}_2\text{O}}{\rightleftharpoons}\ \underset{\substack{\text{Schiff base}\\\text{(not isolated)}}}{\begin{array}{c}\text{RN}\\\|\\\text{CH}\\|\\(\text{CHOH})_n\\|\\\text{CH}_2\text{OH}\end{array}}\ \rightleftharpoons\ \underset{\substack{N\text{-Substituted}\\\text{glycosylamine}}}{\begin{array}{c}\text{RNH}\\|\\\text{HC}\!-\!\!-\!\!-\\|\qquad\;\;\;\text{O}\\(\text{CHOH})_{n-1}\\|\\\text{HC}\!-\!\!-\!\!-\\|\\\text{CH}_2\text{OH}\end{array}}$$

$$\underset{\substack{N\text{-substituted}\\\text{aldosylamine}}}{\begin{array}{c}\text{RNH}\\|\\\text{HC}\!-\!\!-\\|\qquad\text{O}\\(\text{HCOH})_n\\|\\\text{HC}\!-\!\!-\\|\\\text{CH}_2\text{OH}\end{array}}\ \underset{+\text{H}^+}{\rightleftharpoons}\ \underset{\substack{\text{Cation of}\\\text{Schiff base}}}{\left[\begin{array}{c}\text{RNH}\\\|\\\text{CH}\\|\\\text{HCOH}\\|\\(\text{HCOH})_n\\|\\\text{CH}_2\text{OH}\end{array}\right]^+}\ \underset{-\text{H}^+}{\longrightarrow}\ \underset{\text{Enol form}}{\begin{array}{c}\text{RNH}\\|\\\text{CH}\\\|\\\text{COH}\\|\\(\text{HCOH})_n\\|\\\text{CH}_2\text{OH}\end{array}}\ \rightleftharpoons\ \underset{\substack{N\text{-Substituted}\\\text{1-amino-1-deoxy-}\\\text{2-ketose, keto form}}}{\begin{array}{c}\text{RNH}\\|\\\text{CH}_2\\|\\\text{C:O}\\|\\(\text{HCOH})_n\\|\\\text{CH}_2\text{OH}\end{array}}$$

Figure 10.2 Early stages of the Malliard reaction [10].

10.2.2 Evolved Volatile Compounds

The volatile (and steam volatile) compounds which arise when beef and other meats are cooked are obviously very important to flavour and vary both in nature and in quantity according to time and temperature. This is not surprising considering the wide range of temperatures achieved when beef is stewed, grilled or roasted. The volatiles can arise from pyrolysis, sugar dehydration or fragmentation, degradation of amino acids, and inter-reactions of compounds already present or produced during cooking.

Research into their identification has been most thorough. In 1952–1953 many carbonyl compounds, including the significant furfural derivatives, and sulphur compounds (mercaptans, organic sulphides and hydrogen sulphide) were found [4]. Many of these compounds were evolved also from experimental Maillard type reactions between amino acids, including cysteine, and sugars. At first, knowledge of the nature of these volatiles grew slowly. Tonsbeek et al. [13] showed the presence of 4-hydroxy-5-methyl-3(2H)-furanone and 4-hydroxy-2,5-dimethyl-3(2H)-furanone; 2,4,5-trimethyl-3-oxazoline and 3,5-dimethyl-1,2,4-trithiolane [14], and 1-methylthioethane thiol and 2,4,6-trimethyl-perhydro-1,3,5-dithiazine [15] were also identified and shown to contribute significantly to beef aroma. As experimental techniques have been refined, thorough analyses have given identification of many meat volatiles. Comprehensive reviews [16] have named many hundreds of aroma chemicals from cooked beef, though some of these are extremely volatile and some play little part in flavour profile. Some of the more important aroma compounds of meat flavours have been given in a most useful paper by MacLeod [17].

Literature on the volatiles from meats other than beef are not as extensive. Papers on chicken [18], pork [19], ham or bacon [20] and mutton [21] give information which may help in production of a desired specie flavour.

It is outside the scope of this chapter to discuss in detail either the Maillard reaction and meat flavour or the volatiles arising from cooked meats. The literature on both these should be read by those interested in reaction flavours.

10.3 Creating a Process Flavouring

The modern process flavouring is a sophisticated product achieved by mixing in suitable proportions

  (a) the 'basic Maillard' type flavouring product (basic process flavour) (Section 10.3.1) with
  (b) the aroma volatiles (plus any additional flavouring compounds or materials) (Section 10.3.2) and
  (c) other synergists or enhancers, fats or fatty acids or their sources, herbs and spices, etc. (Section 10.3.3)

Depending on the flavour properties required, the reaction flavouring may consist of (a) alone, or the basic flavouring may be blended with either (b) or (c), or a combination of both. The creative flavourist has to be adept in the use of all three parts of the final flavour.

10.3.1 *Basic or Foundation Part of the Flavouring*

This part, produced by non-enzymatic browning, is generally still based on the original patents [5]. An aqueous mixture of amino acids including the important cysteine (though there may be other sources of sulphur) and reducing sugar (though, again, substitutes may be used) is heated under reflux with efficient stirring for a specified time. The reaction conditions are vital; for strong beef products a longer time of heating and perhaps a more concentrated reaction solution is needed, while for chicken and pork products a shorter reaction time, less concentrated solutions and a lower temperature may be appropriate. A reaction mix of pH less than 7 (preferably 2–6) gives better results; a pH above 7 may be more difficult to control as the reaction proceeds rapidly, and the resulting flavour is usually inferior. An increase of pressure for the reaction is sometimes used to increase the temperature and to reduce the time of the reaction.

The key ingredients for the reaction are as follows.

10.3.1.1 *Cysteine or a suitable source of sulphur.* Cysteine reacts readily with reducing sugars. It is now available commercially, as its hydrochloride. The dimer cystine gives generally poorer flavours. Another sulphur-containing amino acid, methionine, gives flavours perhaps reminiscent of potato. Glutathione (γ-L-glutamyl-L-cysteinyl-glycine) also gives good meat flavours, although at present it is too expensive for commercial use.

The importance of sulphur has been recognized for some time, for hydrogen sulphide and other sulphur compounds are evolved when meat is cooked and from model reaction mixes containing cysteine or methionine and sugars, and hydrogen sulphide and xylose will react under increased pressure to produce a strong aroma of beef [4]. The significance of hydrogen sulphide is indicated in a study [22] of the thermal degradation of cysteine and glutathione. A patent describes the reaction of monosaccharides, amino acids and hydrogen sulphide (or a substance giving rise to hydrogen sulphide) [23]. Another describes the reaction of heterocyclic ketones with hydrogen sulphide [24]. Meat flavours are also similarly reported [25] to result from the reaction of 4-hydroxy-5-methyl-3-(2H)-furanone or its thio analogue with hydrogen sulphide; many volatile compounds are identified. There have been many attempts to find a substitute for cysteine hydrochloride, but for practical purposes it remains the most satisfactory sulphur compound in the production of meat flavourings.

10.3.1.2 *The amino nitrogen source.* Whereas a pungent 'raw' flavour can be obtained by heating together cysteine and a reducing sugar, more complete and rounded flavours are obtained if blends of other amino acids are also present. Protein hydrolysates are suitable for this, and are readily available at an economic cost. Protein hydrolysates have been used since the 1930s to improve the flavour of processed foods. They can range from mild flavoured, light-coloured to strong flavoured, deep brown-coloured products. The flavour of every hydrolysate depends on the protein and the conditions of hydrolysis, and is an important contribution to that of the reaction flavour. Virtually any suitable protein may be hydrolysed, by acid or enzyme, to its constituent amino acids. Alkaline hydrolysis is not recommended, for not only are the amino acids arginine and cystine destroyed but inferior tasting hydrolysates result. The protein may be of vegetable or animal origin. Industry generally prefers vegetable hydrolysates (hydrolysed vegetable protein, HVP or hydrolysed plant protein, HPP) prepared by acid digestion of, commonly, the protein from soya, wheat, maize, casein and groundnut. Yeast is usually hydrolysed by enzyme action to give the autolysate, but an acid hydrolysed yeast is used on a small scale. Animal proteins (such as hoof and horn, blood and meat meal,) though quite good for flavour, are not now in general use. Differing amino acid profiles can be obtained from different proteins, as seen in the examples in Table 10.1, taken in part from [4]. Analysis of a lean beef hydrolysate and a yeast autolysate are given for comparison.

It will be seen that wheat gluten hydrolysate contains approximately 38% glutamic acid, whereas other hydrolysates have less than half this amount. This can be useful in giving an 'MSG' flavour effect when no added monosodium glutamate is desired in the food product.

Mixtures of pure amino acids in the proportions of Table 10.1 do not give the same flavours as the corresponding hydrolysates, so the impurities of the hydrolysates must play a part. Besides the aldehydes (including furfural), ketones and sulphur compounds present, α-keto-butyric acid has been isolated from acid-hydrolysed proteins [26], and was thought at one time to be essential to hydrolysate flavour. When this was disproved, it was suggested [27] that α-hydroxy-β-methyl-$\Delta^{\alpha\beta}$-γ-hexenolactone was most important, along with other keto lactones. But this is only a small part of a complicated picture.

Unfortunately some hydrolysate manufacturers are liable to change the protein source without notification, for reasons of economy and availability. This may modify the reaction flavour and it is thus an advantage to have the protein hydrolysate and reaction flavour produced under the same control. Otherwise the flavour and analysis of all hydrolysates used in reaction flavour manufacture should be monitored closely.

Proteins are normally hydrolysed in industry by hydrochloric acid.

Table 10.1 Hydrolysate amino acid analyses as % of total amino acids

Amino acid	Lean beef	Wheat gluten	Soya	Hoof and horn	Yeast (autolysate)
Aspartic acid	7.3	3.8	11.7	9.4	10.1
Threonine	4.4	2.6	3.6	3.8	0.5
Serine	3.7	5.2	4.9	6.8	4.5
Proline	4.3	10.0	5.1	3.2	5.1
Glutamic acid	15.9	38.2	18.5	14.7	11.6
Glycine	4.0	3.7	4.0	7.0	5.4
Alanine	5.6	2.9	4.1	5.1	7.3
Valine	4.3	2.9	5.2	2.1	6.0
Iso-leucine	4.3	1.6	4.6	1.5	4.6
Leucine	7.3	2.8	7.7	3.3	6.9
Tyrosine	2.5	0.2	3.4	1.4	2.9
Phenylalanine	3.0	3.3	5.0	1.6	3.9
Lysine	7.5	1.9	5.8	2.8	7.1
Histidine	2.4	2.0	2.4	0.7	2.2
Arginine	6.0	3.4	7.2	5.8	7.1
Methionine	2.1	1.1	1.2	0.5	1.6

Although the amino acid tryptophan is removed, hydrolysis can be very efficient by use of selective conditions. Sulphuric acid has been used where a low salt content is wanted (an insoluble sulphate can be filtered off after the neutralisation stage) but the taste of the hydrolysate is usually poor. The flavour of the hydrolysate may be influenced by:

(a) the initial nature and purity of the protein
(b) the strength of acid and its proportion to protein
(c) the experimental conditions such as time, temperature and pressure
(d) the grade and method of charcoal treatment, which removes the humins and other impurities and can lower the content of the aromatic amino acids (tyrosine and phenylalanine)
(e) the chemical and method of neutralisation, and final pH of the solution (most liquid hydrolysed plant proteins have a final pH 5–6)
(f) the final concentration or drying process.

The hydrolysate may be concentrated to a paste or dried to a powder. For normal reaction flavour preparation it is more suitable and economical to use the hydrolysate prepared as a liquid of about 40% total solids (of which just under half will be sodium chloride from the neutralisation stage).

Analysis of a typical liquid hydrolysate for reaction flavours should be approximately:

Total solids	40.0%
Sodium chloride	18.5%
Total nitrogen	2.8%
Protein (N × 6.25)	17.5%
Amino nitrogen	2.1%
Weight per ml	1.25
pH	5.6

It is advisable to sieve or filter a liquid hydrolysate before use to remove sediment. This often appears to be deposited on standing, and can be a manufacturing nuisance or even impart a bitter taste to the process flavour. Figure 10.3 summarises the main steps in the commercial production of a protein hydrolysate (from Prendergast [28]).

The need for paste hydrolysates in the food industry is much less than for powder hydrolysates. Dry products are easier to handle and have many uses in processed and convenience foods. Flavour manufacturers often require them for blending with a reaction flavour for particular purposes. Spray-drying is widely used to give a fine powder of uniform size, which will blend well with a spray-dried reaction flavouring. Some manufacturers, however, use a vacuum oven technique which tends to give larger and more irregular particles which have to be graded by sieving; this can be used to advantage, for the powder

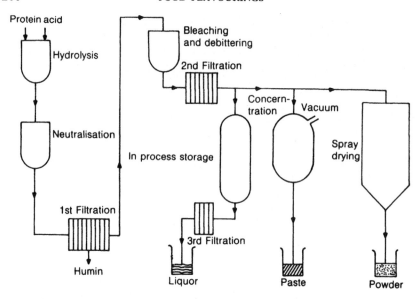

Figure 10.3   Commercial hydrolysate production [28].

residence time at an elevated temperature can impart a roasted or cooked flavour to the hydrolysate from the non-enzymatic browning reaction between amino acids and the impurities.

Autolysates, in particular yeast extract, are made by the action of enzymes on cell components. The resulting low molecular weight compounds include peptides, amino acids and nucleotides. The degree of hydrolysis depends on time, temperature, pH and other process conditions, but is not as complete as with acid hydrolysis of proteins. The yeast *Saccharomyces cerevisiae* gives a strong flavoured autolysate. It has many uses in the food industry, and can be used as the amino nitrogen source in reaction flavour production. The resulting flavourings are of good quality, although they may differ a little in flavour profile. To suit specific purposes, reaction flavourings can be made by employing a mixture of 'acid prepared' hydrolysate and autolysate.

It is most important to choose or develop a protein hydrolysate which is appropriate for the desired reaction flavour.

10.3.1.3 *The sugar component.*   The pentose sugar ribose reacts readily with cysteine and amino acids to give good meat flavours. Soon after its first use [4] it was realised that ribose was very expensive and was not obtainable in commercial quantities, a situation not likely to change for a considerable time. Other sugars were then evaluated including pentoses, hexoses, keto-sugars and disaccharides. It was found that the pentoses gave superior flavours to hexoses, and were more reactive with cysteine and other amino acids, needing less heating time than hexoses to give the meat flavour. Of the four pentoses tried (ribose, xylose, arabinose, lyxose), the best flavour resulted from using ribose; xylose gave the next best flavour, and arabinose and lyxose, though satisfactory as far as flavour was concerned, were not feasible for reasons of availability and cost. Xylose, even in the middle 1950s, was available commercially. Hexoses gave flavours which had many uses. Disaccharides gave poor flavours. Some uses were found for 'replacement' compounds such as furfural and carbonyl compounds. As expected, polysaccharides and starches were ineffective. For practical purposes the most satisfactory sugar is still D-xylose, although other sugars, including glucose, find use in reaction flavours.

The above three ingredients (cysteine or sulphur source, amino nitrogen source, sugar component) will give a range of 'basic' meat flavours. This range can be increased, if desired, by including additional substances such as a fatty material in the mix before reaction [29]. In this case the production of mutton, chicken and pork flavours is facilitated.

10.3.1.4 *Summary of reaction ingredients.* By study of the reaction flavour patents, a comprehensive list of reaction ingredients can be compiled. Table 10.2 gives a list of patented ingredients, taken from MacLeod [17], and indicates the extent of the development of reaction flavourings.

In theory it should be possible for an almost infinite number of reaction flavourings to be produced by permutations of ingredients and reaction conditions. It is necessary, however, to be cautious. From our knowledge of simple Maillard reactions, in which considerable numbers of chemical compounds are produced, it is certain that reactions involving more complicated systems, particularly at higher temperatures and with longer reaction times than normally employed, will yield a larger number of compounds with the possibility that some might be carcinogenic or muta-genic. To decrease this possibility, it is advisable to use simple systems for the reaction, and to avoid reaction overheating. The safety question is discussed in Section 10.5, but as a general principle, it is more satisfactory and ethical for flavour manufacturers to use, as the basic Maillard-type part of the meat flavour, only those reaction flavours which have undergone thorough toxicity tests (as discussed later) or can be said to be harmless by their similarity to the tested products.

10.3.1.5 *Preparations of some basic meat process flavourings.*

Plant equipment: The preferred reaction vessel is of stainless steel, which can be glass-lined. This vessel, of appropriate capacity, should have efficient stirring and be fitted with an efficient reflux condenser. Heating by steam is recommended, for this reduces the danger of over-heating and localised heating. A 'scrubber' or filter system is necessary to remove the pungent volatiles which are produced, and which would otherwise cause environ-

Table 10.2   Some ingredients patented for reaction product meat flavourings [17]

1. **Amino acid/amine sources**
   cysteine, cystine, methionine, alanine, glycine, lysine, arginine, histidine, tryptophan, proline, valine, glutamic acid, glutamine, aspartic acid, glutathione, other S-containing peptides, HVP (groundnut, soybean, wheat/maize gluten), other hydrolysed proteins (casein, egg, fish, blood, liver, bone, collagen), yeast extract, autolysed yeast, meat extract, taurine, pyrrolidone carboxylic acid

2. **Sulphur sources**
   H_2S, cysteine, cystine, methionine, glutathione, thiamin, inorganic sulphides, organic thiols and sulphides, 2-mercaptoethanol derivatives e.g. mercaptoacetaldehyde and/or its dimer 2,5-dihydroxy-1,4-dithiane, 5-hydroxy-3-mercaptopentanone, 3-mercaptopropan-1-ol, 4,5-substituted thiazoles, thiocarbonates, thioamides, 2-mercaptoalkanoic acids/amides, mercaptoalkylamines, aminosulphides, S-acetylmercaptosuccinic acid, vegetable extracts, fermented vegetable juice, yeast extract, autolysed yeast, egg protein, meat extract

3. **Reducing sugar (and derivative) sources**
   ribose, xylose, arabinose, glucose, fructose, lactose, sucrose, ribose-5-phosphate, yeast extract, autolysed yeast, HVP, nucleotides, dextrins, pectins, alginates

4. **α-Diketone and potential α-diketone sources**
   butanedione, pentane-2,3-dione, pyruvaldehyde, pyruvic acid, glyceraldehyde, glyoxal, dihydroxyacetone, α-ketobutyric acid, heptane-3,4-dione-2,5-diacetate, HMFone, HDFone and related derivatives, ascorbic acid, 5-ketogluconic acid, cyclotene, maltol, lactic acid, glycollic acid, malic acid, tartaric acid, protein hydrolysates

5. **Aldehyde sources**
   acetaldehyde, propanal, butanal, methylpropanal, C_5 to C_3 alkanals, HVP

6. **Flavour enhancer sources**
   MSG, IMP, GMP, yeast extract, autolysed yeast, HVP, 2-furfuryl-thioinosine-5'-phosphate, 2-allyloxyinosine-5'-phosphate, 2-(lower alkoxy)inosine-5'-phosphate, 2-benzylthioinosine-5'-phosphate, 4-glucosylgluconic acid i.e. maltobionic acid, cyclotene

7. **pH regulators**
   inorganic acids (e.g. HCl, H_2SO_4, H_3PO_4), organic acids (e.g. succinic, citric, lactic, malic, tartaric, acetic, propanoic), amino acids (e.g. valine, glycine, glutamic acid)

8. **Fat**
   beef, chicken, coconut, triglycerides, fatty acids or esters

9. **Inorganic salts**
   chlorides, phosphates

10. **Spices**

mental problems. A spray-drier or vacuum-oven drier, with screening and packaging equipment, is needed, as only rarely is a reaction flavour liquor required for flavouring purposes. Various models of both these types of drier are commercially available, and it is necessary to experiment to determine the most suitable for each flavour. The usual commercial method is by spray-drying. To facilitate spray-drying, and to reduce hygroscopicity, it is advisable to add a 'carrier' such as a malto-dextrin mixture or cornflour product in the proportions of say 60:40 or 50:50 (flavour solids/carrier). The carrier is stirred into the liquor prior to the liquid entering the drying chamber. A vacuum-oven technique for drying is used by some flavour manufacturers. In this case, as with protein hydrolysates, there can be problems of irregular particle size, and sometimes of over-heating which can give flavour variations. For both methods, an efficient sieving and screening system is necessary. Driers similar to those used for protein hydrolysates are generally suitable.

Examples of basic reaction flavours: The following examples have been adapted from published patents [5, 29]. The ingredient proportions and reaction conditions may be varied for particular requirements.

(a) *Beef flavouring:*

(i) L-Cysteine hydrochloride	80 kg
Protein hydrolysate liquid	1000 kg
(40% total solids,	
strongly-flavoured, lightly bleached,	
from e.g. soya, maize, wheat gluten)	
D-Xylose	100 kg

The hydrolysate liquor is charged to the vessel, followed by (with efficient stirring) the cysteine hydrochloride and xylose. The mix is brought to reflux as quickly as feasible, and heated under reflux with stirring for 3 h. The solution is then cooled as quickly as possible (within 30 min) by circulatory water to below 25°C. The cooled liquor is passed through a fine sieve to remove undesirable matter, and spray-dried to a powder, with flavour solids/carrier in ratio 60:40.

(ii) L-Cysteine hydrochloride	200 kg
Protein hydrolysate	1000 kg
(as in (i))	
D-Glucose	200 kg

The above method is used, but heating under reflux for 6 h. Drying to a powder is the same as for (i).

(b) *Chicken and pork flavourings:* Similar basic reaction flavours may be used for chicken and pork flavours, the characteristic flavour nuances being obtained by the addition of appropriate aroma volatiles, herbs and spices, etc.

(i) L-Cysteine hydrochloride	100 kg
Wheat gluten hydrolysate liquid	1000 kg
(40% total solids, mild-flavoured, well	
bleached)	
D-Xylose	100 kg

The mixture is heated as for beef, but for 60 min only, and carrier added (in ratio 50:50 of flavour solids/carrier), and spray-dried in the same way.

(ii) L-Cysteine hydrochloride	100 kg
Wheat gluten hydrolysate liquid	1000 kg
(as for (i))	
D-Xylose	60 kg

Lard (or chicken fat)	500 kg
Water	3000 kg

The mixture is heated as for beef with efficient stirring for 2 h. After the reaction the mixture is cooled to 80–85°C and glyceryl monostearate (33–40% monoglyceride) (20 kg) added, with vigorous stirring. A carrier as above (500 kg) is added and the stirring continued for 30 min. The homogenised mixture is then spray-dried.

   (c) *Smoked ham flavouring:*

L-Cysteine hydrochloride	60 kg
Soya hydrolysate liquid (40% total	1000 kg
solids: medium-flavoured, medium-bleached)	
D-glucose	40 kg
Smoke solution	2000 kg

The mixture is heated as for beef for 4 h, and spray-dried after additions of the usual carrier (ratio flavour solids/carrier 60:40). Smoke solution is prepared by heating oak sawdust in a kiln, condensing the vapours, distilling the condensate under reduced pressure to remove the more volatile, unwanted compounds, discarding tarry materials, and diluting the remainder with water. Smoke solutions are available commercially which should conform to the proposed Council of Europe specifications. Variation in quantity and strength of the smoke solution will give a lightly smoked to heavily smoked flavour.

   (d) *Roast mutton/lamb flavour:*

L-Cysteine hydrochloride	110 kg
Protein hydrolysate liquid (as for (a.i))	1500 kg
D-Xylose	140 kg
Oleic acid	100 kg

The mixture is heated as for beef for 2 h, and spray-dried (ratio flavour solids/carrier 50:50), making certain of efficient dispersion of flavour and carrier.

Many more examples of basic reaction flavourings have been published, nearly all in patents. A proportion of these include the use of expensive individual amino acids and sugars, which would result in flavours much too costly for use in process foods. None of these examples are given here as they are not considered suitable for commercial use at present. It is worthwhile; however, to be aware of the availability and price of these materials in case their use does become feasible.

10.3.2 *The Aroma Volatiles*

In this section, all the aroma chemicals, whether of high or low volatility, are classed as 'volatiles'. From the hundreds of volatile materials which have been reported as being evolved on cooking various meats, it is obviously important to select those of advantage in giving the desired additional meat aroma. Many compounds cannot be used as they are too volatile for use in a flavouring material, and some may even have a negative effect.

The potency of a large number of these compounds is remarkable, so they can only be employed in minute amounts. Because of this, it is usually necessary to prepare a solution of the compounds in a solvent such as isopropyl alcohol (propylene glycol can be used but may cause the formation of lumps when dispersed on the carrier powder, and thus give difficulties in the blending and handling of the final flavour). Additional flavouring compounds or materials may, if soluble, be dissolved in the volatiles/isopropyl alcohol solution, or dispersed separately onto the carrier if a liquid, or blended with the final flavour if a powder.

10.3.2.1 *Beef flavour.* Lawrie [30] gives a short list of six volatile compounds of significance for meat flavour:

pyrrolo-[1,2-*a*]-pyrazine
4-acetyl-2-methyl pyrimidine
4-hydroxy-5-methyl-3-(2*H*)-furanone
4,6-dimethyl-2,3,5,7-tetrathiaoctane
2-alkylthiophene
3,5-dimethyl-1,2,4-trithiolane

These can be used as a base for a 'volatile mix'. Perusal of the literature [16] will suggest a large number of additional compounds. A selection of some which could contribute to meat aroma, taken from MacLeod [17], is given in Table 10.3.

In addition there are many groups of compounds which can give 'lift', 'bite', 'mellowness', 'roundness', etc. to a variety of flavours, and these are well known in the flavour industry. Of importance in this respect are aldehydes and ketones, furans and derivatives, sulphur compounds, thiophens and derivatives, and the important pyrazines. The number of compounds which can be used in the volatiles 'cocktail' depends on many factors. Many are not commercially available (though the number being marketed is growing rapidly) and it is thus very helpful to have the services of skilled synthetic chemists. Many substances are too expensive for flavouring use even in the minute amounts necessary. Some may be unstable, but may respond to stabilising methods. It could be a lengthy procedure to select and blend together a mixture of the feasible volatile compounds to make a flavour volatiles solution to complement the 'basic reaction' flavouring. However, this should be within the capabilities of the good flavourist.

For practical purposes in preparing the final, complete flavouring blend, it is preferable to have the volatiles dispersed on a 'carrier'. The most suitable

Table 10.3   Volatile compounds for beef aroma [17]

2-Methyl-3-(methylthio) furan
3-Hydroxybutane-2-thiol
3-Mercaptopentanone
2-Methylcylopentanone
3-Methylcyclopentanone
2-Methylfuran-3-thiol
2,5-Dimethyl furan-3-thiol
2-Methyl-3-(methyl thio)furan
2-Methyl-3-(methyldithio)furan
2-Methyl-4,5-dihydrofuran-3-thiol
bis-(2-Methyl-3-furyl)trisulphide
bis-(2-Methyl-3-furyl)disulphide
bis-(2-Methyl-3-furyl)tetrasulphide
5-Methyl furfurylthiol
bis-(2-Furfuryl)disulphide
2-Methyl-thiophen-3-thiol
2-Methyl-4,5-dihydrothiophen-3-thiol
2-Methyl-4,5-dihydrothiophen-4-thiol
2-Methyl-4-hydroxy-2,3-dihydrothiophen-3-thiol
2,5-Dimethyl-4-hydroxy-2,3-dihydrothiophen-3-one
2,5-Dimethyl-2,4-dihydroxy-2,3-dihydrothiophen-3-one
2-Methyl-tetrahydrothiophen-4-thiol
2-Methyl-1,3-dithiolane
1,2,4-Trithiolane
3-Methyl-1,2,4-trithiane
2,4,6-Trimethyl-1,3,5-trithiane
1,2,3-Trithiacyclo-hept-5-ene
Thiazole
2,4-Dimethyl-5-ethylthiazole
thialdine or 2,4,6-trimethyi-1,3,5-dithiazine
5-Acetylthiazole
2,5-Dimethyl-4-hydroxy-3(2H)-furanone
2,5-Dimethyl-3(2H)-furanone
5-Methyl-2-furaldehyde

carrier is a carbohydrate such as lactose, glucose or maltose derivatives (e.g. malto-dextrins). Care has to be taken or some volatiles may be lost. The solution of blended volatiles is mixed with the carrier, sieved to ensure even particle size and the powder stored in sealed containers for the final mixing stage. Similarly, volatile powder mixes can be prepared for other types of meat flavours such as those described below.

10.3.2.2 *Chicken flavour.*   The aroma from cooked chicken is less pungent than that from beef, as a result not only of the nature of the meat, but of the usual methods of cooking. The phospholipids play a significant role, with their content of arachidonic, linolenic and linoleic acids. The characteristic aroma of chicken has been said to be largely due to *cis*-4-decenal, *trans*-2-*cis*-5-undecadienal and *trans*-2-*cis*-4-*trans*-7-tridecatrienal; other unsaturated compounds such as *cis*-3-nonenal, 2-*trans*-dodecenal, 2-*trans*-5-*cis*-undecadienal,

2,4-dodecadienal and 2,6-dodecadienal have been found in cooked chicken and can be used along with related compounds [31, 32]. A patent [33] also describes how certain aliphatic aldehydes (11–17 carbon atoms and 2–4 olefinic bonds) can impart a chicken-type aroma. In addition to these compounds, the chicken 'addition mix' could contain 2,5-dimethyl-2,5-dihydroxy-1,4-dithiane [34] and the compound 'sulphurol' (4-methyl-5-thiazole ethanol) [35]. (Sulphurol is readily available and is useful in most types of meat flavours.)

Some pyrazines (e.g. 2,5-dimethyl pyrazine, 2-methyl pyrazine and 2,3,5-trimethyl pyrazine) will give a cooked or roasted note and 2,4,6-trimethyl-1,3,5-trithiane and thialdine (2,4,6-trimethyl-5,6-dihydro-1,3,5-dithiazine) [36] also help. 2-Butyl furan, dipropyl sulphide, 2-methyl furan and 2-pentyl furan and other simple sulphides and furan compounds may be of benefit. Dimethyl trisulphide [37] will intensify the overall effect of a chicken volatiles blend.

As they are reared today, most chickens when cooked have somewhat weak and bland flavours. A recent paper [38] gives the volatile composition of the roasted guinea hen, which is said to be of stronger flavour, and finds little qualitative difference from that of roasted chicken. This supports the view that imitation chicken flavours can be improved and intensified considerably by use of a carefully blended volatiles component in the correct proportions.

10.3.2.3 *Pork flavour.* This is a delicate flavour, as is chicken. Many of the chicken and beef aroma chemicals can be used if blended carefully. Thiamine, which has been used in reaction flavours [39], is present in pork in larger quantities than in beef; when heated it will give 3-mercaptopropanol, 3-acetyl-3-mercaptopropanol and 4-methyl-5-vinylthiazole, which are useful aroma chemicals for pork flavour mixes. The aroma profile can be improved by including trace quantities of 3,5-di-(2-methylpropyl)-1,2,4-trithiolane and 2,4,6-tri-(2-methylpropyl)tetrahydro-1,3,5-dithiazine [40]; 2-decenoic acid, 2-heptenoic acid, 2-nonenoic acid, 2,3-diethyl-5-methyl pyrazine, 2-acetyl-3-ethyl pyrazine, 4-methylthiazole and 2,4,5-trimethyloxazole are helpful ingredients. 2-Pyridine methanethiol is claimed to contribute to roast pork flavour [41].

10.3.2.4 *Smoked ham flavour.* Ham flavourings can be prepared ranging from 'green bacon' to 'heavily smoked ham'. This is done by the addition of other basic reaction flavourings (e.g. beef, pork in varying amounts) to the smoked ham reaction flavouring, and by the use of selected aroma materials. 3,5-Di-(-2-methylpropyl)-1,2,4-trithiolane and 2,4,6-tri-(2-methylpropyl) tetrahydro-1,3,5-dithiazine, which possess fried bacon odours [17, 40] would, if available, improve the lightly smoked flavour. Propionyl furan, acetyl furan and various guaiacols (e.g. 4-ethyl guaiacol, 4-methyl guaiacol, guaiacol phenyl acetate), 5-methyl furfural and minute traces of 2,5-dimethyl pyrazine,

2-vinyl pyrazine and 2,3,5-trimethyl pyrazine can supplement and intensify the more smoked flavours.

10.3.2.5 *Roast mutton/lamb flavour.* In some countries lamb is not a favoured food, though in the United Kingdom and many other countries, it is well-liked. Its particular flavour is derived from the lipids, and is more specific than that of other meats. Many compounds contribute to this flavour, including 4-methyloctanoic acid, 4-methylnonanoic acid, and other branched-chain acids [42]. Three compounds thought to occur only in mutton volatiles are bis-(mercaptomethyl)sulphide, 3,6-dimethyl-1,2,4,5-tetrathiane and 1,2,3,5,6-pentathiepane (lenthionine) [43]. 2-Ethyl-3,6-dimethyl pyrazine and 2-pentylpyridine [44] contribute to the fatty odour. Pentanal, hexanal, heptanal, octanal, nonanal, 2-octenal, 2,4-heptadienal (and isomer), 2,4-decadienal (and isomer), 2-nonanone, 2-dodecanone, 2-tridecanone, γ-octalactone [45], and 3,5-dimethyl-1,2,4-trithiolane and thialdine (2,4,6-trimethyl-1,3,5-dithiazine) [43] can also be important. Small amounts of the aroma chemicals used in other meat flavours, in particular sulphurol and 4-hydroxy-5-methyl-3-(2H)-furanone, can again be used for 'rounding off' the lamb flavour.

Aroma mixes for other less common meat flavours can be composed by careful and painstaking flavour methods. Examples of these may include veal, kidney, liver, venison, pheasant, duck and oxtail. Appropriate basic reaction flavours for these mixes have to be selected.

Flavouring volatiles, which may be classed as 'nature identical' or said to arise from cooking of meats, normally constitute only a small part of a reaction flavour. Nevertheless it is necessary to check that they satisfy 'safety in use' requirements.

10.3.3 *Enhancers, Fats, Herbs and Spices*

10.3.3.1 *Enhancers.* This part of the reaction flavour can include any of the ingredients of part 6 of Table 10.2, but the substances usually considered for this function are monosodium glutamate, disodium 5'-inosinate and disodium 5'-guanylate.

Flavour enhancers or flavour potentiators are substances which have little flavour themselves, but when added to foodstuffs have the property of enhancing or intensifying the flavour of the product. They can also sometimes suppress undesirable taints, and can improve the important mouthfeel of a food product.

The isolation, identification and initial development work of these three enhancers was in Asia and commercial production of them is still concentrated there. They are used extensively in the world-wide flavour industry.

(a) *Monosodium glutamate:* This is the monosodium salt of L-glutamic acid.

It has been known to have flavour enhancing properties for about 80 years, but it was only in the early 1950s that its use became more widespread, as means of preparing it became commercially feasible and the introduction and popularity of convenience and process foods caused a real need. Large scale production is now mainly by fermentation.

Monosodium glutamate (or MSG as it now universally known) has in recent years suffered from over-use; it has in some cases been added indiscriminately to prepared foods, sometimes to no flavour advantage. In some countries it even appears on the table as an extra condiment. When food is over-dosed with MSG the taste becomes unpleasant and the chemical can then be recognised by many people. Not surprisingly, a public reaction against MSG has developed and tales of 'Chinese restaurant syndrome' and 'MSG migraine' have become common. With help from the media, it has become an additive to be avoided. More recently, MSG has been employed with much more discretion, and, used sensibly, is accepted as a useful food ingredient.

There is no synergistic effect by incorporating MSG in a reaction flavour, for the flavour impact in the final food product is the same whether the MSG effect is obtained from MSG in a reaction flavour or from MSG added separately to the food. The main reason for blending MSG into the reaction flavour is by request of the flavour user, who may for various reasons (such as the need to reduce 'operative stages') not wish to add the MSG himself.

(b) *Disodium 5'-inosinate and disodium 5'-guanylate:* Both these compounds are ribonucleotides, having closely related chemical structures (Figure 10.4). Of the 5'-ribonucleotides which exist in nature, only these two show marked flavour enhancement properties. They are normally marketed for flavour work as their disodium salts.

Disodium inosinate and disodium guanylate are the sodium salts of 5'-inosinic acid (inosine-5'-mono phosphoric acid) (IMP) and 5'-guanylic acid (guanosine-5'-monophosphoric acid) (GMP). The flavour effects of these compounds have been known for sometime, but they were not produced commercially until 1959. They each have similar flavour enhancing properties, although the guanylate is over twice as effective as the inosinate. They are available as individual compounds, but are commonly produced and sold as a

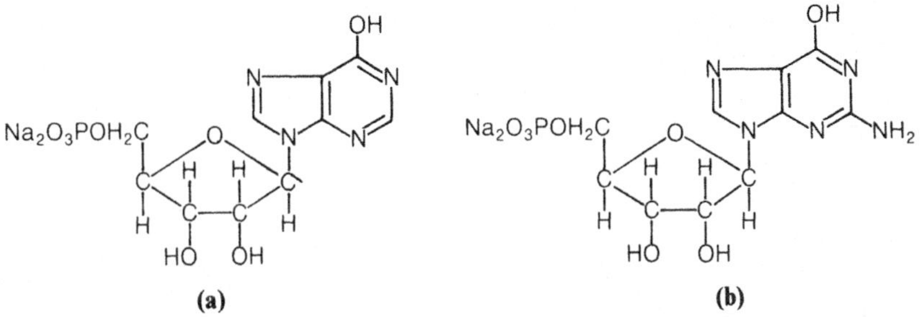

Figure 10.4   Structure of (a) disodium 5'-inosinate and (b) disodium 5'-guanylate.

50:50 blend of inosinate/guanylate. They are now prepared by fermentation or enzymatic hydrolysis, and find a wide use, not only in savoury products. The inosinate/guanylate blend is much more potent than MSG, perhaps 50–100 times stronger. When used together with MSG in food products, there is a definite synergistic effect, one recommended ratio being in the region of 1 part ribonucleotide to 50 parts MSG. A decided advantage can be gained by including the inosinate/guanylate mix in a reaction flavour blend. Very little is needed. At say, 0.1–0.5%, the flavouring effect of the reaction flavouring appears to be increased.

10.3.3.2 *Fats.* The reaction flavour can in some cases be improved by including a small amount of the appropriate fat or perhaps fatty acids. Some fats can be obtained as powders, but in cases of a liquid fat (e.g. rendered chicken fat), it is necessary to disperse the fat onto a carrier such as lactose, with the help where needed of a flowing agent (calcium silicate is quite effective). The carriers can absorb up to approximately 10% of their weight of liquid fat.

10.3.3.3 *Herbs and spices.* The addition of herb and spice preparations means that a wider and more extensive range of reaction flavours can be made. The very ancient trade in spices and herbs has become very specialised, and today a selection of different herb and spice products is produced in many countries by specialised techniques [46]. They are available as dried and ground spices, oleoresins and essential oils. Ground spices can suffer microbial contamination, though this may be reduced considerably by irradiation which is permitted in some countries. (Sterilising by treating with ethylene oxide and propylene oxide is not entirely satisfactory and may slightly alter the flavour.)

Oleoresins, prepared by solvent extraction, also contain non-volatile constituents and are obtained as pastes or viscous oils. It is not easy to convert them into free-flowing powders, necessitating a high proportion of 'carrier'. The essential oils are perhaps the most satisfactory, being readily absorbed into carriers. They are sometimes distilled 'in the field' i.e. near to where harvested, and can vary in quality. More standard products can be prepared in factories with good technical equipment, by or on behalf of flavour companies. As all herb and spice preparations may be subject to adulteration, it is always necessary to check the quality of all essential oils before acceptance and use [46, 47]. If ground herbs or spices are used, it is necessary to examine them microbiologically.

Because of their potency, most spice and herb essential oils are blended as solutions in a convenient solvent such as isopropyl alcohol, with strengths from 5 to 50%. When a desired blend has been obtained to suit a particular flavour need, the mix of spices/herbs is dispersed onto a suitable carrier such as lactose or malto-dextrin product and sieved and stored in sealed containers before the final reaction flavour mixing stage.

The following are examples of herb and spice blends for specialised flavourings:

(a) *Spiced braised beef:* Oils of bay, pepper, cinnamon, thyme, onion, lovage, parsley, marjoram, bouquet garni mix.

(b) *Beef and onion:* Very little herbs and spices but a little garlic oil, plus a little pepper and mustard oils. It is an advantage to include onion powder and perhaps garlic powder, though their microbiological count can be high, depending partly on their country of origin.

(c) *Barbecued beef:* Many recipes exist, but good blends can be obtained from oils of black pepper, nutmeg, clove, coriander, thyme, with extracts of ginger and capsicium and some onion.

(d) *Curried beef:* Many ingredients go into a good curry formulation, the types of which depend on their origin. Usually a general simple formulation is best, and can be prepared from oils of coriander, cumin, cardamon, clove, cinnamon, ginger and nutmeg. The bright yellow finely ground turmeric powder not only contributes to the flavour but gives a distinctive colour to the final reaction flavour.

(e) *Chicken with stuffing:* Any good spice stuffing mix is suitable, and could be based on oils of celery, parsley, sage Dalmatian, nutmeg, bay, thyme, cumin and fenugreek.

(f) *Pork with stuffing:* As with chicken, many recipes exist, and suitable examples could be based on oils of sage Dalmatian, thyme, nutmeg, mace, ginger, black pepper, celery and parsley. Separately, a good apple flavour helps to round-off a roast pork flavour, though in this case the stuffing note should be reduced considerably.

(g) *Smoked ham:* To be kept to a minimum, but could include oils of cloves, ginger, rosemary, basil and bay. Addition of a honey flavour in trace quantities helps to obtain a smoother flavour.

(h) *Lamb/mutton:* This flavour is definitely helped by a mix of oils of rosemary, mint and possibly traces of parsley, cinnamon and ginger.

Many publications give examples of spice and herb mixes [46] and the flavourist can use these and the suggestions above for preliminary work in developing the most suitable mix for each requirement. As a general principle it is important that the herb/spice mix should be adjusted so that in the food product as eaten, it is at a level which complements and not dominates the flavour effect.

Unless added to the basic reaction mixture before heating for a special purpose (e.g. chicken fat for chicken flavour) it is advisable to blend aroma volatiles, enhancers, fats, herbs and spices with the post-reaction, dried, basic Maillard reaction product. This is because they may lose part of their flavour character or effect by being volatile, unstable to heat, or by taking part themselves in the Maillard reaction.

10.3.4 *Compounding the Complete Process Flavouring*

It is quite possible that a competent flavour company may be able to offer several hundred different 'complete' reaction flavourings, most of which have been developed to satisfy the needs of individual food products. It is unlikely, however, (for toxicity reasons, discussed later) that there will be more than say half a dozen basic (or Maillard reaction) flavourings used in these. The variety and number of flavourings is produced by modifications in the remaining parts of the final flavourings.

Final mixing of the three main parts of the flavouring (the basic or foundation part, the aroma volatiles, and the enhancers, fats, herbs and spices) should be in a dry, well-ventilated and easily cleaned room. Operatives should be well protected with special clothing and nose/mouth masks, as there will be dust in the air during mixing and packing. There are satisfactory types of powder mixer/blenders, sieves and packaging equipment available. A work-sheet should accompany each flavour preparation. The aroma volatiles and herbs and spice mixes are made separately and dispersed onto the carrier (e.g. lactose). The required amounts of basic/Maillard flavouring, aroma volatiles and herbs and spice mixes, perhaps enhancers and fats are added to the mixer/blender, and after a suitable period of mixing the whole is sieved and packed. It is very important to keep all operations in a very dry atmosphere, as the flavourings are hygroscopic. An outline example of a typical flavouring mix may be:

Basic/Maillard reaction flavouring	75%
Aroma volatiles, as solution in isopropyl alcohol, dispersed on lactose at 10%	15.0%
Herb/spice blend, 20% in isopropyl alcohol, dispersed on lactose at 10%	9.7%
Sodium ribonucleotides (mix of inosinate/guanylate)	0.3%

These flavourings are stable for many months if kept sealed in air-tight containers at ambient or below ambient (say below 20°C) temperatures. Raised temperatures during storage should be avoided as some of the volatile materials may evaporate and thus upset the balance of the flavouring. Temperatures below 0°C, should also be avoided as this may cause the formation of powder 'lumps' and general deterioration of fine powder structure.

10.4 Applications of Process Flavourings

Process flavourings can make a general improvement to many savoury food products such as dried or liquid soups, gravies and stock cubes, canned foods, casseroles, meat pies, savoury biscuits, snacks, spreads, whole 'instant' and factory-prepared meals, new protein foods and meat analogues. The method of adding the flavour can be most important for the success of the food product.

If the processed or convenience food is to undergo much cooking after purchase by the consumer (which is unusual), it may be necessary to 'overload' the reaction flavouring with the volatile materials to allow for loss. Most of these foods as purchased are already cooked and merely need heating.

Flavourings can be added to a food product by:

(a) simple mixing with other ingredients in the final blending of a convenience food (Section 10.4.1)
(b) application to the outside of a food, such as with snack products (e.g. crisps); this is sometimes referred to as 'dusting' (Section 10.4.2)
(c) addition to a food during its formation, such as in the preparation of textured proteins or some snack products; this is sometimes referred to as 'internal' flavouring (Section 10.4.3)

10.4.1   Simple Mixing

Examples of this can be soup mixes (dry materials in a sealed pack, or a wet mix in a metal can), cooked prepared meals, gravies, meat pies and pastes, or canned products such as baked beans with, e.g. meat or bacon flavour. In most cases there is little heat processing after the addition of the reaction flavouring. If there is much heating there may be need to use a partially reacted flavouring, or even include a proportion of flavour precursors with a reduced amount of this partially reacted flavouring. Special care has to be taken if reaction flavourings are sealed in a canned product which is then heated at high temperatures; this may cause black sulphur-staining of the can lining. This can be reduced considerably, however, by choice of a suitable lacquered can, by choice of a suitable (adapted) reaction flavouring, and by adhering strictly to recommended processing conditions.

10.4.2   Application to Outside ('Dusting')

The change in eating habits over the last 30–40 years has been marked not only by the growth of convenience foods but by the remarkable growth of the snack food market.

In the United Kingdom, perhaps the most popular of these snack foods are potato crisps and extruded snacks. For the latter, engineering techniques have produced products which can be flavoured before extrusion, but in addition flavouring can be dusted on the outside at a later stage if required. With potato crisps, the flavouring may be dusted on by established methods. The dusting powder on the product outside has the advantage of an immediate flavour impact in the mouth.

Dusting powders are made by mixing savoury or meat flavourings with 'extenders' which may be mixtures of very fine rusks, flours, autolysates and hydrolysed vegetable proteins, and perhaps a free flowing agent. As soon as potato crisps or other snack products are taken out of the frying oil, dusting powders are sprinkled over them. During cooling, the powder is absorbed into the product, and, theoretically, sets fast in the surface fat. In practice, a proportion of the dusting powder often shakes free in the packet. The 'complete' reaction flavourings are quite satisfactory for incorporation into dusting powders; as no processing in the product is involved, there will be little loss of volatile materials.

Because of their mainly polar nature, reaction flavours do not dissolve or disperse well into oils. Nevertheless there have been attempts to spray flavour dispersions on to the cooked crisps or snack products. This has been only partially successful.

10.4.3 Addition During Product Manufacture

This differs from simple mixing (where perhaps only moderate heat and other physical treatment may be involved) as sometimes severe physical and chemical processing can be part of the manufacture of a food product. We are concerned here with two types of food, the meat analogue or simulated meat (a proteinaceous food product made to be fairly close to various meats in texture and nutritive value) and the extruded snack product.

10.4.3.1 The meat analogue.

The two main methods of preparation of meat analogues are thermoplastic extrusion and the spinning process, usually employing soya protein. Extrusion methods start with a mix (say 5–10% moisture) of defatted protein with additives (e.g. starch, vegetable oils, flavouring, colouring) at pH 6.5–7.0. This dough is heated and forced under increased pressure through an extruder and a restricted orifice. Its temperature rises considerably for a short time because of the heater and friction. As the dough leaves the orifice and emerges to atmospheric pressure, it loses water and expands to a 'foam' which can be cut into required sizes. The flavouring has to be added before extrusion, and thus a part of the volatiles may disappear with the loss of water. Reaction flavourings for this type of product should preferably be made from the basic reaction plus enhancers alone. The short time of heating should not cause much extra browning reaction to occur.

Spun protein foods have been developed which are similar in appearance to natural meats. The process, briefly, involves passing a viscous solution of protein at a high pH through 'spinnerettes' of minute diameter (0.002–0.006 inches) into a salt/acid bath. Filaments are formed which are washed and cut into appropriate lengths. By the use of a binder to hold the fibres together, the final food product can be formed. It is not easy to flavour spun protein. If the flavour is added at the 'dope' stage (i.e. before passage through the spinnerettes), the high pH may destroy part of the flavouring, and some of the flavouring will be leached out by the later washing stage. This latter difficulty can be overcome by using excess flavouring (which can be expensive) in combination with a prior flavouring of the binder. It is important to add flavouring and perhaps some flavouring precursors before spinning, so that flavouring may be formed and 'locked' in the fibres in situ during processing. This will ensure that flavour sensation lasts during chewing of the food.

Extruded textured vegetable proteins which are intended to replace a small part of the meat content of a food product do not always need additional flavour, as they can absorb it from the meat and other ingredients. However if they are present at more 'than about 25%, it does help to have flavoured material. Spun protein analogues must have adequate internal and perhaps also external flavouring. In both types of product the object of the flavouring is not only to give the required flavour, but to mask the sometimes inherent poor (perhaps 'beany') taste of the treated protein.

Some other protein foods which have been developed as meat alternatives do not necessarily undergo the vigorous extrusion or spinning techniques. Flavouring of such foods (e.g. single cell or microbial protein) is more straight forward, and can be near the end of the process, without much loss of flavour quality.

10.4.3.2 *The extruded snack product.* The snack industry uses extrusion technology to manufacture many of its products, creating new interest by variations in shape and flavour. Both sweet and savoury flavourings can be used, but in the United Kingdom beef, beef and onion, chicken, bacon and cheese flavourings remain very popular. These snacks have very little nutritive value, but their popularity does illustrate the gradual change in eating habits. They are not primarily based on protein, but are made from a variety of raw materials which can include corn products, salt, wheat flour and starches. Processing techniques are critical to the product. Two main methods used are direct expansion (collet) and indirect expansion (pellet; this is expanded during subsequent deep-fat frying). A third method (co-extrusion filled tubes, direct expansion with simultaneous 'cream-filling') is under development, but this gives a rather specialised type of product. Both internal and external flavouring, and a combination of both, can be employed in snack production by these two methods. It is even possible to have one flavouring (e.g. beef) on the inside, and one lightly dusted on the outside (e.g. onion). With internal flavouring, thorough mixing is essential and over-heating should be avoided because of possible flavour loss.

It is usually advisable to 'tailor-make' flavourings, with the possible inclusion of a proportion of flavour precursors to allow for subsequent heating of the snack product. The pellet-snacks, before expansion, have a longer shelf-life. If desired they can be sold, already flavoured internally with, e.g. basic beef/chicken/pork reaction flavours, to other manufacturers who could then add dusting powder flavours of their choice.

Although reaction flavourings find a wide use in a variety of applications, it is essential that the flavourist and the food technologist cooperate closely when a food product is being developed. This cooperation will ensure that any processing or manufacturing conditions of the food product will work to advantage in producing a product with a well-balanced flavour.

10.5 The Safety Question

Reaction flavourings are a comparatively recent part of the flavour industry, and were introduced at a time of much research into the safety of all materials connected with food and food additives. It was only to be expected that the toxicity factor of these flavourings should be investigated thoroughly. Protein hydrolysates have also been the subject of much interest because, as well as being important reaction flavour ingredients, they are used widely in food products as flavouring ingredients in their own right.

10.5.1 Safety of Protein Hydrolysates

There was no doubt about the non-toxic status of protein hydrolysates until, recently [48], some chlorine-containing organic compounds were found as contaminants, and identified as chlorohydrins. They were shown to be most probably derived from the lipids of the protein to be hydrolysed. The amounts of these contaminants were very low, but the industry reacted quickly. Hydrolysate manufacturers combined research and knowledge to develop methods for eliminating or greatly reducing them in manufactured protein hydrolysates.

The compounds under investigation may be classified as follows.

(a) *Monols:* These consist of monochloropropanols (MCPs), e.g. 1-chloro-propanol, 2-chloropropanol and 3-chloropropanol, and dichloropropanols (DCPs), e.g. 2,3-dichloropropan-1-ol and 1,3-dichloropropan-2-ol.

(b) *Diols:* Examples are 2-chloro-1,3-propandiol and 3-chloro-1,2-propandiol (MCPDs).

Analytical methods have been devised to examine protein hydrolysate contaminants. The determination of 1,3-dichloropropan-2-ol [49] is now a routine test available for hydrolysates, the detection limit being 50 ppb DCP. Hydrolysed proteins produced by most manufacturers are now guaranteed to contain amounts of DCP below this detectable level. The levels of the other contaminants have been reduced considerably, and their determination in protein hydrolysates, with permissible limits, is under active and constant review. A recent article [50] gives a summary of attitudes to these contaminants, which have been discussed both by the European Parliament and by individual governments. It has to be stressed that the amounts of these contaminants in hydrolysed proteins have always been extremely small. When the hydrolysate as now prepared is part of a reaction flavouring used in a food product at say from 0.05 to 0.5% by weight, it is logical to state that there should not be any potential hazard from this source.

10.5.2 Safety of Reaction Flavourings

The earliest meat reaction flavourings were evaluated most thoroughly for possible toxicity, and the results submitted to the Food and Drug Administration (FDA), Washington, DC, as a part of an appraisal of protein hydrolysates in food [51]. Short-term and long-term animal feeding studies on beef and chicken flavours (cysteine/HVP/xylose or glucose) and smoked ham flavour (cysteine/HVP/xylose/liquid smoke preparation) showed no toxic effects.

In the last 10–12 years there has been much discussion and research concerning mutagens and carcinogens which arise on cooking protein (muscle) foods at high temperatures [52]. The possibility that mutagens could be found in most meat foods was supported by an investigation into twenty commercially available meat flavourings [53] consisting of natural and nature identical components. The natural materials gave positive mutation results and the nature identical negative results (with one exception in each case). In a discussion on the mutagenic activity of heated foods Powrie et al. [54] mentioned two potent mutagens, coded Try-P-1 and Try-P-2 which arise from pyrolysis of tryptophan, and two, Glu-P-1 and Glu-P-2, which arise from glutamic acid. It is extremely unlikely that any of these mutagens could arise at the temperatures (up to 105°C) used for most reaction flavours, and the ones from tryptophan could not arise in any case because this amino acid has been removed during the processing of the commercial protein hydrolysates.

The Council of Europe is discussing the proposal of I.O.F.I. that the maximum reaction temperature should be 180°C for no longer than 15 min (with correspondingly longer times at lower temperatures). This temperature was proposed because it is reached in the household kitchen in frying fat. This recommendation does not specify the reaction time in reaching this high temperature, nor the cooling period, which, unless very sophisticated reaction equipment is used, must be of consequence. It is to be hoped that a future Council directive will clarify this ambiguity for any further flavour development, for temperature and time parameters are of importance for safety considerations.

Most of the reaction flavours now in general use in the United Kingdom and in Europe are similar to or are developments of the first meat reaction flavours. As a class they have been tested more thoroughly than any other type of flavour. They can be said to be non-toxic, and have at no time showed any cause for concern.

10.6   Process Flavourings in the Future

As the food industry continues to expand its range of processed and convenience foods the flavour industry will expand to accommodate this. It would be unwise to speculate on possible growth of meat analogues and substitute foods (previous forecasts have not been very accurate) as this depends on so many factors involving politics, economics and whether or not novel or substitute foods will be accepted and enjoyed by the potential consumer.

Reaction flavouring production has increased markedly, and continues to increase, since they were first produced in the United Kingdom in the late 1950s, and they now account for about 65% by weight of the total flavours made in Europe (excluding solvents and natural foods, adjuncts, etc.) [55]. Because of the necessary lengthy and involved toxicity trials, it is unlikely that many new basic reaction systems will be developed. It can be forecasted, however, that a large number of reaction flavourings will be produced which rely on the ingredients (with known safety data) added after thermal processing for their variety of flavour effects.

For this reason, it is apt that the Council of Europe guidelines should consider dividing the ingredients into two broad classes:

(a) ingredients usually added prior to thermal processing
(b) ingredients added only after thermal processing

When the final revised version of the guidelines becomes European law, industry will be obliged to comply with its requirements. Because of the necessity of confidentiality in the flavour industry, there will have to be a deep sense of trust and free exchange of information between the Council of Europe and the individual companies producing and using reaction flavourings.

A large part of the flavour industry is currently engaged on the development and application of reaction flavourings. The extent of their growth over the past 30 years has been remarkable, and if the public taste continues to prefer savoury tastes to sweet tastes, and ready-made or processed foods, there is no doubt there will be an increasing demand for the flavours which originated in the cooking of simple amino acids and sugars. The amount of research time, energy and money that has been and will be expended on the production of imitation meat flavours shows the great need for this type of flavouring in our developing society.

Acknowledgements

I am grateful to Dr. W.T. Little and Unilever Research for permission to describe the flavour research I carried out at Colworth House, Sharnbrook, Bedford, and to Mr. G.A. Cowell and the directors of H.E. Stringer Limited for their support in writing this chapter. I acknowledge the help of Mr. W.J. Scott and Mrs. Frances Rule in assembling and typing this text.

Appendices

Appendix I: Composition of Lemon and Orange Oils

Lemon Oil, Italian

Obtained by pressing peels of *Citrus limon* L., N.L. Burman, a lemon species grown in Italy. Pale yellow to pale greenish-yellow liquid with a characteristic lemon peel odour.

Evaporation residue	1.6–3.6%
Acid number max	1.4%
Carbonyl compounds content, calculated as citral	3.0–5.0%

Main components of lemon peel oils are terpenes and composition depends on variety grown and country of origin, e.g. in American oils,

(+)-Limonene	65%
beta-Pinene	8–10%
gamma-Terpinene	8–10%

The characteristic odour of lemon oil, which is different from that of other citrus oil is caused to a large extent by the two citrals:

Neral and geranial	less than 3%
	(but can be as high as 10%)

The esters neryl and geranyl acetate contribute to the initial aroma impression and changes with increasing alpha-terpineol and terpinen-4-ol content. These alcohols are, in part, artefacts formed under the influence of the citric acid present in the juice during the production of the oil.

Orange Oil, Sweet

Obtained by cold pressing the peels of the orange species, *Citrus* sinesis L. Osbeck or, for Guinea type oils, of the varieties *limoviridis* A. Chevalier and *djalonis* A. Chevalier. Yellow to reddish-yellow liquid with the characteristic odour of orange peel; the oil may become cloudy when chilled. The physical properties of the oil depend on the variety and origin:

	Italy	Guinea	Brazil
Evaporation residue (%)	1.6–3.5	1.0–3.2	2.0–3.6
Carbonyl compounds content (as decanal) (%)	0.9–2.2	1.8–3.1	1.4–3.1

Sweet orange oil is produced in many countries in combination with orange juice, e.g. United States, Brazil, Israel and Italy. Terpene hydrocarbon content, mainly (+)-limonene always > 90%. Various oils differ in oxygen-containing compounds. The aroma is determined by the aldehydes, mainly octanal, decanal, and both citrals and esters, mainly octyl and neryl acetate. The sesquiterpene aldehydes alpha- and beta-sinensal, which also occur in other citrus oils, although in lower concentrations, contribute particularly to the specific sweet orange aroma.

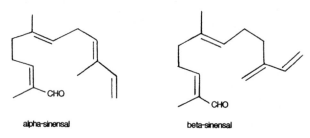

alpha-sinensal   beta-sinensal

Appendix II: Botanical Classification of Fruits

Citrus

Fleshy, edible fruit, segmented into sections and surrounded by a thick skin which contains essential oil characteristic of the fruit: orange; lemon; lime; tangerine/ mandarin; grapefruit; citron; bergamot; kumquat.

Berries

Generally small fruits (exception kiwi) with pulpy edible part which has many seeds and generally juicy: blackcurrant; blackberry; blueberry; boysenberry; cranberry; gooseberry; huckleberry; kiwifruit; loganberry; raspberry; redcurrant; strawberry.

Fruits Containing Stones

Contains a single (normally large) stone which is the seed surrounded by the fleshy, edible pericarp: apricot; cherry; damson; peach; plum.

Tropical Fruits

Avocado; banana; date; fig; mango; pineapple; pomegranate.

Pomes

Body of fruit is formed by swelling of recepticle surrounding seed capsule: apple; pear; quince.

Melons

Large fruit with either soft pulpy or cellular flesh containing many seeds: water; honeydew; musk; canteloup.

Grapes

Small, usually very sweet fruits growing in clusters on vines with both seed-containing and seed-less varieties. There are many varieties of both red and white.

Appendix III: International organization of the Flavour Industry (I.O.F.I.) Guidelines for the Production and Labelling of Process Flavourings

Introduction

Process flavourings are produced by heating raw materials which are foodstuffs or constituents of foodstuffs in similarity with the cooking of food.

The most practicable way to characterise process flavourings is by their starting materials and processing conditions, since the resulting composition is extremely complex, being analogous to the composition of cooked foods. They are produced every day by the housewife in the kitchen, by the food industry during food processing and by the flavour industry.

The member associations of I.O.F.I. have adopted the following Guidelines in order to assure the food industry and the ultimate consumer of food of the quality, safety and compliance with legislation of process flavourings.

1. Scope

1.1. These Guidelines deal with thermal process flavourings, they do not apply to foods, flavouring extracts, defined flavouring substances or mixtures of flavouring substances and flavour enhancers.

1.2. These Guidelines define those raw materials and process conditions which are similar to the cooking of food and which give process flavourings that are admissible without further evaluation.

2. Definition

A thermal process flavouring is a product prepared for its flavouring properties by heating food ingredients and/or ingredients which are permitted for use in foodstuffs or in process flavourings.

3. Basic Standards of Good Manufacturing Practice

The chapter 3 of the Code of Practice for the Flavour Industry is also applicable to process flavourings.

4. Production of Process Flavourings

Process flavourings shall comply with national legislation and shall also conform to the following:

4.1. Raw materials for process flavourings. Raw materials for process flavourings shall consist of one or more of the following:

4.1.1. A protein nitrogen source:

- Protein nitrogen containing foods (meat, poultry, eggs, dairy products, fish, seafood, cereals, vegetable products, fruits, yeasts) and their extracts
- hydrolysis products of the above, autolysed yeasts, peptides, amino acids and/or their salts.

Table A.1 Materials used in procession

Herbs and spices and their extracts
Water
Thiamine and its hydrochloric acid salt
Ascorbic acid
Citric acid
Lactic acid
Fumaric acid
Malic acid
Succinic acid
Tartaric acid
The sodium, potassium, calcium, magnesium and ammonium salts of the above acids
Guanylic acid and inosinic acid and its sodium, potassium and calcium salts
Inositol
Sodium, potassium- and ammoniumsulfides, hydrosulfides and polysulfides
Lecithine
Acids, bases and salts as pH regulators:
   Acetic acid, hydrochloric acid, phosphoric acid, sulphuric acid
   Sodium, potassium, calcium and ammonium hydroxide
   The salts of the above acids and bases
Polymethylsiloxane as antifoaming agent (not participating in the process)

4.1.2. A carbohydrate source:
- foods containing carbohydrates (cereals, vegetable products and fruits) and their extracts
- mono-, di- and polysaccharides (sugars, dextrins, starches and edible gums)

4.1.3. A fat or fatty acid source:
- foods containing fats and oils
- edible fats and oils from animal, marine or vegetable origin
- hydrogenated, transesterified and/or fractionated fats and oils
- hydrolysis products of the above.

4.1.4. Materials listed in Table A.1

4.2. Ingredients of process flavourings

4.2.1. Natural flavourings, natural and nature identical flavouring substances and flavour enhancers as defined in the I.O.F.I. Code of Practice for the flavour industry.

4.2.2. Process flavour adjuncts. Suitable carriers, antioxidants, preserving agents, emulsifiers, stabilisers and anticaking agents listed in the lists of flavour adjuncts in Annex II of the I.O.F.I. Code of Practice for the flavour industry.

4.3. Preparation of process flavourings. Process flavourings are prepared by processing together raw materials listed under 4.1.1. and 4.1.2. with the possible addition of one or more of the materials listed under 4.1.3. and 4.1.4.

4.3.1. The product temperature during processing shall not exceed 180°C.

4.3.2. The processing time shall not exceed 1/4 hour at 180°C, with correspondingly longer times at lower temperatures.

4.3.3. The pH during processing shall not exceed 8.

4.3.4. Flavourings, flavouring substances and flavour enhancers (4.2.1) and process flavour adjuncts (4.2.2) shall only be added after processing is completed.

4.4. General requirements for process flavourings.

4.4.1. Process flavourings shall be prepared in accordance with the General Principles of Food Hygiene (CAC/Vol A-Ed. 2 (1985)) recommended by the Codex Alimentarius Commission.

4.4.2. The restrictive list of natural and nature-identical flavouring substances of the I.O.F.I. Code of Practice for the Flavour Industry applies also to process flavourings.

5. Labelling

The labelling of process flavourings shall comply with national legislation.

5.1. Adequate information shall be provided to enable the food manufacturer to observe the legal requirements for his products.

5.2. The name and address of the manufacturer or the distributor of the process flavouring shall be shown on the label.

5.3. Process flavour adjuncts have to be declared only in case they have a technological function in the finished food.

References

Chapter 1

1. Dorland and Rogers, in *The Fragrance and Flavour Industry*. Wayne E. Dorland, Box 264, Mendham, NJ 07945 (1977).
2. M.E. Porter, in *Competitive Strategies*, Collier Macmillan, London (1980).
3. F. Rosengarden Jr. Spices, in *Tropical Products Institute Conference Proceedings*, T.P.I. (1973).
4. *Food Manufacture: Ingredients Survey*, Morgan-Grampian, London (1972).
5. Hall and Oser, FEMA GRAS Substances 3, *Food Technol.* **19**(2) (1965) 151.
6. Hall and Oser, FEMA GRAS Substances, Vol. 4, *Food Technol.* **24**(5) (1970) 25; Oser and Hall, FEMA GRAS Substances, Vol. 5, *Food Technol.* **26**(5)(1972) 35; Vol. 6, **27**(1) (1973) 64; Vol. 7, **27**(11)(1973) 56; Vol. 8, **28**(9)(1974) 76; Vol. 9, **29**(9)(1975) 70; Vol. 10, **31**(1)(1977) 65; Vol. 11, **32**(2) (1978) 60; Vol. 12, **33**(7) (1979) 65; Vol. 13, **38**(10) (1984) 65.
7. Council Directive on the Approximation of the Laws of the Member States Relating to Flavourings for Use in Foodstuffs and to Source Materials for their Production (88/388/EEC), *Official Journal of the European Communities* No. L 184/61 (15.7.88).
8. Council Decision on the Establishment, by the Commission of an Inventory of the Source Materials and Substances used in the Preparation of Flavourings. *Official Journal of the European Communities* No. L 184/67 (15.7.88).
9. *Code of Practice for the Flavour Industry*, 2nd edn. (1985), 1st amendment (1988), IOFI, 8 rue Charles-Humbert, CH-1205 Geneva, Switzerland.
10. Stofberg and Stoffelsma, Consumption of flavouring materials as food ingredients and food additives, *Perfumer and Flavourist* **5**(7) (1980) 19; Stofberg, Consumption ratio and food predominance of flavouring materials, 1st series, *Perfumer and Flavourist* **8**(3) (1983) 61; Stofberg and Grundschober, 2nd series, *Perfumer and Flavourist* **9**(4) (1984) 53; 3rd series, *Perfumer and Flavourist* **12**(4) (1987) 27.
11. Rulis et al., FDA's priority-based assessment of food additives,
    1. Preliminary results, *Regul. Toxicol. Pharmacol.* **4** (1984) 37;
    2. General toxicity parameters, *Regul. Toxicol. Pharmacol.* **5** (1985) 152;
    3. Specific toxicity parameters, *Regul. Toxicol. Pharmacol.* **6** (1986) 181;
    Rulis et al., in *A Codex Flavour Priority Ranking System*, Codex Committee on Food Additives, 19th session (1987).

Chapter 2

1. B.M. Lawrence, *Perfumer and Flavourist*, Vol. 10(5) Oct/Nov (1985) 1.
2. *Essential Oils and Oleoresins: A Study of Selected Producers and Major Markets*, International Trade Centre UNCTAD/GATT (1986).
3. R.L. Swaine, *Perfumer and Flavourist*, Vol. 13(6) Dec (1988) 1.

Chapter 3

1. N. Shaath and P. Griffin, Frontier of flavor, in *Developments in Food Science*, Vol. 17 ed. G. Charalambous, Elsevier, Amsterdam (1988) pp. 89–108.
2. Analytical Methods Committee of Chemical Society, Essential Oils Subcommittee, *The Analyst* (London) **109** (1984) 1343.

3. Felton Worldwide Limited, *Templar Essential Oils, part I. CO$_2$ Extract Applications*, Felton, Bilton Road, Bletchley, MK1 1HP, UK (1989).
4. Analytical Methods Committee, *The Analyst* (London) **113** (1987) 1125.
5. Pfizer—C.A.L., *Technical Data Sheet—Celery Seed Oil*, C.A.L., Grasse, France (1986).
6. D.A. Moyler, in *Proceedings Conference Amer. Chem. Soc.* Crete, ed. G. Charalambous, Elsevier, Amsterdam (1989).
7. D.A. Moyler and H.B. Heath, Flavors and fragrances world perspective, in *Developments in Food Science*, Vol. 18, eds. B.M. Lawrence, B.D. Mookherjee and B.J. Willis, Elsevier, Amsterdam (1988).
8. B.M. Lawrance, *Perfumer and Flavorist* **10** (1985) 1.
9. D. McHale, W.A. Laurie and J.B. Sheridan, *Flavour and Fragrance J.* **4** (1989) 9.
10. H.B. Heath, in *Source Book of Flavors*, AVI (1982).
11. J.-Q. Cu, F. Perineau, M. Delmas and A. Gaset, in *Proceedings ICEOFF*, New Delhi, India **4** (1989) 89.
12. D.A. Moyler, in *Proceedings ICEOFF*, New Delhi, India (1989) Abstract 84.
13. J.P. Calame and R. Steiner, in *Theory and Practice in Supercritical Fluid Technology*, eds. M. Hirata and T. Ishikawa, Tokyo Metropolitan Univ. (1987) 277–318.
14. S.N. Naik, H. Lentz and R.C. Maheshwari, *Fluid Phase Equilibria* **49** (1989) 115.
15. S.N. Naik, R.C. Maheshwari and A.K. Gupta, in *Proceedings ICEOFF*, New Delhi, India (1989) Abstract 202.
16. J.A. Pickett, J. Coates and F.R. Sharpe, *Chem. Ind.* (1975) 7.
17. P.A.P. Liddle and P. de Smedt, *Parfum Cosmet Arome* (1981) 42.
18. A.H. Rose, ed., *Alcoholic Beverages, Economic Microbiology*, Vol. 1 Academic Press (1977).
19. Felton Worldwide Limited, *Templar Oils, part VIII. Vanilla Extracts*, Felton, Bilton Road, Bletchley, MK1 1HP (1989).
20. Gebruder Wollenhaupt, *Real Vanilla*, Vanille Import-Export GmBH, 2057 Reinbek, Hamburg W.G. (undated).
21. B.E.M.A. British Essence Manufacture Assoc., *Guidelines on Allowable Solvent Residues in Foods* and *Natural Flavouring Definition*, BEMA, 6 Catherine Street, London, WE2B 5JJ.
22. Felton Worldwide Limited, *Templar Oils, part III, Molecular Distillation*, Felton, Bilton Road, Bletchley, MK1 1HP, U.K. (1989).
23. D.A. Moyler, Chem. Ind. **18** (1988) 660.
24. D.A. Moyler, in *Distilled Beverage Flavour*, eds. J.R. Piggott and A. Paterson Ellis Horwood (1988).
25. D.A. Moyler, in *Theory and Practice in Supercritical Fluid Technology*, eds. M. Hirata and T. Ishikawa, Tokyo Metropolitan Univ. (1987) 319–341.
26. D.S.J. Gardner, *Chem. Ind.* (London) **12** (1982) 402.
27. D.A. Moyler, *Perfumer and Flavorist* **9** (1984) 109.
28. H. Brogle, *Chem. Ind.* **12** (1982) 385.
29. K.U. Sankar, *J. Food Sci. Agric.* **48** (1989) 48.
30. N. Gopalakrishnan, P.P.V. Shanti and C.D. Narayanan, *J. Food Sci. Agric.* **39** (1988) 3.
31. C.C. Chen, M.C. Kuo and C.T. Ho, *J. Agric. Food Chem.* **34** (1986) 477.
32. C.C. Chen and C.T. Ho, *J. Agric. Food Chem.* **36** (1988) 322.

Chapter 4

1. R.M. Smock and A.M. Neubert, in *Apples and Apple Products*, Interscience Publications, London (1950).
2. J.P. Van Buren and W.B. Robinson, *J. Agric. Food Sci.* **17** (1969) 772.
3. K. Wucherpfennig, P. Possmann and E. Kettern, *Flussiges Obst.* **39** (1972) 388.
4. J.F. Kefford, H.A. McKenzie and P.C.O. Thompson, *J. Sci. Food Agric.* **10** (1959) 51.
5. *The Soft Drinks Regulations*, H.M.S.O., London, Statutory Instrument (1964) No. 760.
6. V.L.S. Charley, *Food Technol.* **18** (1963) 33.
7. *The Fruit Juices and Nectars Regulations*, H.M.S.O., London, Statutory Instrument.
8. *Methods of Analysis*, International Federation of Fruit Juice Producers, Zug, Switzerland, Various Dates.
9. *Official Methods of Analysis*, 14th edn. A.O.A.C., Arlington, VA.

10. J.B. Redd, C.M. Hendrix and D.L. Hendrix, in *Quality Control Manual for Processing Plants*, Intercit Inc. (1988).
11. *R.S.K. Values, The Complete Manual*, Association of German Fruit Juice Industry, Bonn, Flussiges Obst. (1988).

Chapter 5

1. S. van Straten and H. Maarse, ed., in *Volatile Compounds in Food*, 5th edn., TNO, Zeist (1988) Suppl. 5.
2. Chemical Source Association, *Flavor and Fragrance Materials—1889*, Allured Publishing Co., Wheaton (1989).
3. S. Arctander, in *Perfume and Flavor Chemicals*, Vols. 1, 2, published by the author, Montclair, NJ (1969).
4. G. Fenaroli, in *Fenaroli's Handbook of Flavor Ingredients*, Vols. 1, 2, CRC Press, Boca Raton, FL (1975).
5. E. Gildemeister and F. Hoffmann, eds., in *Die ätherischen Oele*, Akademie Verlag, Berlin (1968).
6. E. Guenther, in *The Essential Oils*, Vols. 1–6, D. van Nostrand, New York (1952).
7. G.G. Birch and M.G. Lindley, in *Developments in Food Flavours*, Elsevier, Amsterdam (1986).
8. E. Ziegler, in *Die natürlichen und künstlichen Aromen*, A Hüthig Verl., Heidelberg (1982).
9. H.B. Heath, in *Source Book of Flavours*, AVI Publishing Co., Westport, CT (1981).
10. H.B. Heath and G.A. Reineccius, in *Flavor Chemistry and Technology*, AVI Publishing Co., Westport, CT (1986).
11. H.B. Heath, in *Flavor Technology: Profiles, Products, Application*, AVI Publishing Co., Westport, CT (1978).
12. F.M. Clydesdale, ed., in *Critical Reviews in Foods Science and Nutrition*, Vol. 28, CRC Press, Boca Raton, FL (1989).
13. P.Z. Bedoukian, *Perfumer and Flavorist* **14** (1989) 2.
14. Chemical Abstracts Service (editorial office), *Chemical Abstracts*, Vol. 111, American Chemical Society, Columbus, OH (1989).
15. H.W. Schultz, E.A. Day and L.M. Libbey, in *The Chemistry and Physiology of Flavors*, AVI Publishing, Westport, CT (1967) 331.
16. R. Tressl, F. Drawert, W. Heimann and R. Emberger, *Z. Lebensm. -Unters Forsch.* **144** (1970) 4.
17. H.-D. Belitz and W. Grosch, in *Lehrbuch der Lebensmittelchemie*, Springer, Berlin (1987) pp. 18–19.
18. O. Wallach, in *Terpene und Campher*, Vit., Leipzig (1914).
19. L. Ruzicka, A. Eschenmoser and H. Heusser, *Experientia* **9** (1953) 357.
20. T.W.G. Solomons, Biosynthesis of isoprenoids, in *Organic Chemistry*, John Wiley, New York (1988) pp. 1080–1085.
21. P. Schreier, in *Chromatographic Studies of Biogenesis of Plant Volatiles*, A Hüthig Verlag, Heidelberg (1984) pp. 52–126.
22. P. Manitto, The isoprenoids, in *Biosynthesis of Natural Products*, Ellis Horwood, Chichester (1981) pp. 215–297.
22a. R. Croteau, *Chem. Rev.* **87** (1987) 929.
23. L.C. Maillard, *C.R. Seances Acad. Sci.* **164** (1912) 66.
24. C. Eriksson, ed., in *Maillard Reaction in Food. Chemical Physiological and Technological Aspects*, Pergamon Press, Oxford (1981).
25. G.R. Waller and M.S. Feather, eds., in *The Maillard Reaction in Foods and Nutrition*, ACS *Symp. Series* **215**, American Chem. Soc., Washington DC (1983).
26. F. Ledl, G. Fritul, H. Hiebl, O. Pachmays and T. Severin, Degradation of Maillard products, in *Amino-Carbonyl Reactions in Food and Biological Systems*, eds. M. Fujimaki, M. Namiki and H. Kato, Elsevier, Amsterdam (1986) p. 173.
27. F.D. Mills and J.E. Hodge, *Carbohydrate Res.* **51** (1976) 9.
28. Y. Houminer, *Thermal Degradation of Carbohydrates in Molecular Structure and Function of Food Carbohydrate*, eds. G.G. Birch and L.F. Green, Applied Science Publishers, London (1973) pp. 133–154.

29. H.-D. Belitz and W. Grosch, in *Lehrbuch der Lebensmittelchemie*, Springer, Berlin (1987) pp. 206–272.
30. M. Moo-Young, ed., in *Comprehensive Biotechnology*, Vols. 1–4, Pergamon Press, Oxford (1985).
31. H.-J. Rehm and G. Reed, in *Biotechnology*, Vols. 1–8, VCH Verlagsges., Weinheim (1989).
32. N.A. Porter, L.S. Lehmann, B.A. Weber and K.J. Smith, *J. Am. Chem. Soc.* **103** (1981) 6447.
33. J.M.H. Bemelmans and M.C. ten Noever de Brauw, Analysis of off-flavours in food, in *Proc. Int. Symp. Aroma Research*, Pudoc, Wageningen (1975) pp. 85–93.
34. H. Maarse and R. Belz, in *Isolation, Separation and Identification of Volatile Compounds in Aroma Research*, Akademie Verlag, Berlin (1985).
35. P. Schreier, in *Techniques of Analysis, in Chromatographic Studies of Biogenesis of Plant Volatiles*, A. Hüthig Verlag, Heidelberg (1984) pp. 1–51.
36. P. Schreier, ed., *Analysis of Volatiles*, de Gruyter, Berlin (1984).
37. A.J. MacLeod, *Instrumental Methods in Food Analysis*, Elek Science, London (1973).
38. J. Adda, ed., in *Development in Food Science, 10. Progress in Flavour Research*, Elsevier, Amsterdam (1985).
39. R.G. Berger, S. Nitz and P. Schreier, eds., *Topics in Flavour Research*, H. Eichhorn Verlag, Marzling-Hangenham (1985).
40. L.B. Kier, *J. Pharm. Sci.* **61** (1972) 394.
41. J.E. Amoore, Odor theory and odor classification in fragrance chemistry, in *The Science of the Sense of Smell*, ed. E.T. Theimer, Academic Press, New York (1982) pp. 27–76.
42. R. Tressl, D. Bahri and K.-H. Engel, Lipid oxidation in fruits and vegetables, *Am. Chem. Soc. Symp. Series* **170** (1981) 213.
43. J.M. Kerr and C.J. Suckling, *Tetrahedron Lett.* **29** (1988) 5545.
44. P.Z. Bedoukian, in *Perfumery and Flavoring Synthetics*, Allured Publishing, Wheaton (1986) pp. 16–18.
45. P. Dubs and R. Stüssi, *Synthesis* (1976) 696.
46. W.G. Jennings and R. Tressl, *Chem. Mikrobiol. Technol. Lebensm.* **3** (1974) 52.
47. A. Ijima and K. Takahashi, *Chem. Pharm. Bull.* **21** (1973) 215.
48. P. Oberhänsli, CH Pat. 576 416 (assigned to Givaudan, 1976).
49. A. Nissen, W. Rebafka and W. Aquila, Eur. Pat. 21 074 (assigned to BASF, 1981).
50. D. Helminger and P. Naegeli, US Pat. 3 943 177 (assigned to Givaudan, 1976).
50a. G. Bucchi and H. Wuest, *J. Am. Chem. Soc.* **96** (1974) 7573.
51. D.A. Andrews and C.N. Hindley, US Pat. 3 994 936 (assigned to Hoffmann-La Roche, 1973).
52. M.W. Breuninger, Eur. Pat. 51 229 (assigned to Hoffmann-La Roche, 1980).
53. R. Croteau, M. Felton and R. Ronald, *Arch. Biochem. Biophys.* **200** (1980) 534.
54. R. Noyori and M. Kitamura, Enantioselective catalysis with metal complexes, in *Modern Synthetic Methods 1989*, ed. R. Scheffold, Springer, Berlin (1989) p. 181.
55. J. Fleischer, K. Bauer and R. Hopp, DOS 2 109 456 (assigned to Haarman and Reimer, 1972).
56. P. Schreier, in *Chromatographic Studies of Biogenesis of Plant Volatiles*, A. Hüthig Verlag, Heidelberg (1984) p. 120.
57. R. Tressl, M. Holzer and M. Apetz, Biogenesis of volatiles in fruits and vegetables, in *Aroma Research*, eds. H. Maarse and P.J. Groenen, Pudoc, Wageningen (1975) p. 41.
58. P.Z. Bedoukian, *Perfumery and Flavoring Synthetics*, Allured Publishing, Wheaton (1986) pp. 267–282.
58a. H. Mayer and O. Isler, in *Total Synthesis in Carotenoids*, ed. O. Isler, Birkhaüser, Basel (1971) pp. 328–575.
59. G. Saucy and R. Marbeth, *Helv. Chim. Acta* **50** (1967) 1158.
60. P.G. Bay, US Pat. 3,062,874 (assigned to Glidden, 1962).
61. P. Schreier, in *Chromatographic Studies of Biogenesis of Plant Volatiles*, A. Hüthig Verlag, Heidelberg (1984) pp. 84–97.
62. G. Peine, *Chem. Ber.* **17** (1884) 2109.
63. H.D. Belitz and W. Grosch, in *Lehrbuch der Lebensmittelchemie*, Springer, Berlin (1987) p. 297.
64. N. Elming and N. Clauson-Kaas, *Acta Chem. Scand.* **6** (1952) 867.
65. *The Givaudan Index*, Givaudan-Delawanna Inc., New York (1962) p. 252.
65a. G.P. Rizzi, *Food Reviews Internat.* **4**(3) (1988) 375.
66. W. Baltes and G. Bochmann, *Z. Lebensm. -Unters Forsch.* **184** (1987) 485.
67. H. Masuda, M. Yoshida and T. Shibomoto, *J. Agric. Food Chem.* **29** (1981) 944.

68. JA Pat. 6 0258 168 (assigned to Koei Chem. Ind. KK, 1984).
69. I. Flament, DOS 3 048 031 (assigned to Firmenich, 1979).
70. H. Masuda and S. Mihora, *J. Agric. Food Chem.* **34** (1986) 377.
71. B.D. Mookherjee, C. Ciacino, E.A. Karoll and M.H. Vock, DOS 2 166 323 (assigned to IFF, 1974).
72. H.J. Wild, *Chem. Ind.* (1988) 580.
73. J.E. Hodge, *Am. Soc. Brew. Proc.* (1963) 84.
74. U. Huber and H.J. Wild, US Pats. 4 181 666 and 4 208 338 (assigned to Givaudan, 1977).
75. H.J. Wild, WO Pat. 87'03 287 (assigned to Givaudan, 1986).
76. G.A.M. Ouweland and H.G. Peer, Austrian Pat. 340 748 (assigned to Unilever, 1977).
77. H. Sulser, J. de Pizzol and W. Büchi, *J. Food Sci.* **32** (1967) 611.
78. G.J. Hartman, J.T. Carlin, J.D. Scheide and C.-T. Ho, *J. Agric. Food Chem.* **32** (1984) 1015.
79. W. Roedel and U. Hempel, *Die Nahrung* **18** (1974) 133.
80. S.-S. Hwang, J.T. Carlin, Y. Bao, G.J. Hartman and C.-T. Ho, *J. Agric. Food Chem.* **34** (1986) 538.
81. G. MacLeod, The scientific and technological basis of meat flavors, in *Developments in Food Flavors*, eds. G.G. Birch and M.G. Lindley, Elsevier, London (1986) pp. 191–223.
82. R.G. Buttery, W.F. Haddon, R.M. Seifert and J.G. Turnbaugh, *J. Agric. Food Chem.* **32** (1984) 674.
83. H. Sugimoto and K. Hirai, *Heterocycles* **26** (1987) 13.
84. W.J. Evers, H.H. Heinsohn, B.J. Mayers and A. Sanderson, *Am. Chem. Soc. Symp. Ser.* **26** (1976) 184.
85. U. Huber, US Pat. 4 285 984 (assigned to Givaudan, 1977).
86. G. Ohloff, Chemie der Geruchs- u. Geschmacksstoffe, in *Fortschritte der chemischen Forschung*, ed. F. Boschke, Springer, Berlin (1969) pp. 185–251.
87. R. Tressl, D. Bahri and K.-H. Engel, *J. Agric. Food Chem.* **30** (1982) 89.
88. R. Kaiser and C. Nussbaumer, *Helv. Chim. Acta* **73**(1) (1990) 133–139.
89. A. von der Gen, *Perf. Cosm. Sav. France* **2** (1972) 356.
89a. M. Hamberg, *J. Am. Oil Chem. Soc.* **66** (1989) 1445.
90. T.-L. Ho, *Synth. Commun.* **4** (1974) 265.
91. P. Canonne, J. Plamondon and M. Akssira, *Tetrahedron* **44** (1988) 2903.
92. R.J. Parry, E.E. Mizusawa, I.C. Chiu, M.V. Naidu and M. Ricciardone, *J. Am. Chem. Soc.* **107** (1985) 2585.
93. C.T. Pedersen and J. Becher, eds., in *Developments in the Organic Chemistry of Sulfur*, Gorden and Breach, New York (1988).
94. R.H. Cagan and R.M. Care, eds., in *Biochemistry of Taste and Olfaction*, Academic Press, New York (1981).
95. H.E. Muerman, *Perfumer and Flavorist* **14** (1989) 1.

Chapter 6

1. J. Bergstrom and E. Hultman, *J. Am. Med. Assoc.* **221** (1972) 999; National academy of Sciences—National Research Council, Water deprivation and performance of athletes, *Nutr. Rev.* **32** (1974) 314.
2. K. Bauer and D. Gerbe, in *Common Fragrance and Flavour Materials—Preparation, Properties and Uses*, VCH, Cambridge (1985) 117.
3. H.B. Heath, *Source Book of Flavours*, AVI, Westport, CT (now part of van Nostrand Reinhold, New York)
4. D.K. Tressler and M.A. Joslyn, *Fruit and Vegetable Juice Processing Technology* (2nd ed.), AVI, Westport, CT (now part of van Nostrand Reinhold, New York).
5. L. Schroeter, in *Sulphur Dioxide Applications in Food, Beverages and Pharmaceuticals*, Pergaman Press, Oxford (1966); E.W. Taylor, *The Examination of Waters and Water Supplies*, J.A. Churchill, London (1958).
6. P.W. Harmer, private communication.
7. *The Soft Drinks Regulations 1964—Composition and Labelling*, S.I. No. 760, H.M.S.O., London (1971).

8. D. Hicks in *Non-Carbonated Fruit Juices and Fruit Beverages* (ed. Hicks), Blackie, Glasgow, 1990.

Chapter 7

1. *Butter Regulations*, Statutory Instrument 1966, No. 1074, H.M.S.O., London (1966).
2. *Food Labelling Regulations*, Statutory Instrument 1984, No. 1305, H.M.S.O., London (1984).
3. *Skuses Complete Confectioner*, 13th edn., W.J. Bush & Co., Ltd., London (1957).
4. J. Ames, Maillard Reaction, *Food Manufacture*, November (1989) 6.

Chapter 8

1. *The Bread and Flour Regulations*, Statutory Instrument 1963, No. 1435, H.M.S.O., London (1963).
2. Kent-Jones and Amos, in *Modern Cereal Chemistry*, The Northern Publishing Co., London (1967).
3. *Flavor Chemistry, Advances in Chemistry*, Series 56, The American Chemical Society (1969).
4. A.R. Daniel, in *Up to Date Confectionery*, Maclaren & Sons, London (1947).
5. P. Wade, in *Biscuits Cookies and Crackers*, Vol. 1, Elsevier Applied Science, London (1988).
6. S.A. Matz, in *Cookie and Cracker Technology*, AVI, Westport, CT. (now part of Van Nostrand Reinhold, New York) (1968).
7. D.J.R. Manley, in *Technology of Biscuits and Cookies*, Ellis Horwood, Chichester (1983).
8. R.J. Taylor, in *Food Additives*, John Wiley, New York (1980).
9. J. Ames, Maillard Reaction, *Food Manufacture*, November (1989) 61.

Chapter 9

1. D.A. Forss, *J. Dairy Res.* **46** (1979) 691.
2. J.P. Dumont and J. Adda, in *Proc. 2nd Weurmann Flavour Research Symposium*, eds. D.G. Land and H. Nursten (1979) Chapter 21.
3. P.M. Boon, A.R. Keen and N.J. Walker, *New Zealand J. Dairy Sci. Technol.* **11** (1976) 189.
4. *Food and Drugs—Composition and Labelling: The Cream Regulations* S.I. 752 (1970).
5. Anon, *Dairy Industries International* **54**(8) (1989) 19.
6. K.J. Burgess, in *Food Industries Manual*, 22nd edn., ed. M.D. Ranken, Blackie, Glasgow (1988) Chapter 3.
7. T. van Eijk, *Dragoco Rep.* **3** (1986) 63.
8. M. Byrne, *Food Proc.* August (1989) 19.
9. A.Y. Tamine and H.C. Deeth, *J. Food Protection* **43**(12) (1980) 939.
10. *Code of Practice for the Composition and Labelling of Yogurt*, Dairy Trade Federation (1983).
11. *Volatile Compounds in Food, Qualitative Data*, 5th edn., *Quantitative Data*. Vol. 2, eds. H. Maarse and C.A. Visscher, Nutrition and Food Research TNO, Zeist, The Netherlands (1983).
12. V.M. Marshall, *Perfumer and Flavorist* **7** (1982) 27.
13. E.J. Mann. *Dairy Industries International* **54**(9) (1989) 9.
14. *Food and Drugs—Composition and Labelling: The Butter Regulations*, S.I. 1074 (1966).
15. P.B. McNulty, *Proc. SCI Symposium*, September (1975) 236.
16. J.E. Kinsella, *Food Technol.* **29**(5) (1975) 82.
17. D.A. Forss, G. Urbach and W. Stark, in *Proc. 17th Int. Dairy Congress* **C:2** (1966) 211.
18. E.H. Ramshaw, *Australian J. Dairy Technol.* **29**(3) (1974) 110.
19. G. Christian, in *Cheese and Cheese-Making*, Macdonald, London (1977).
20. V. Marquis and P. Haskell, in *The Cheese Book*, Simon & Shuster, New York (1985).
21. B.A. Law, in *Proc. Dairy Symposium*, Elsevier (1984) Chapter 7.
22. B.A. Law, *Dairy Science Abstracts* **43**(3) (1981) 143.
23. B.A. Law, *Perfumer and Flavorist* **7** (1982) 9.

24. K.H. Ney, in *The Quality of Foods and Beverages—Proc. 2nd Int. Flavour Conf. Athens*, ed. G. Charalambous, Academic Press, New York (1981) p. 389.
25. G. Christian, in *World Guide to Cheese*, Ebury Press, London (1984).
26. C. Karahadian, D.B. Josephson and R.C. Lindsay, *J. Agric. Food Chem.* **33** (1985) 339.
27. J. Fernandez-Salguero, A. Marcos, M. Alcala and M.A. Esteban, *J. Dairy Res.* **56** (1989) 141.
28. D.J. Manning, *Dairy Industries International* **43**(4) (1978) 37.
29. J.F. Horwood, G.T. Lloyd and W. Stark, *Aust. J. Dairy Technol.* **36**(1) (1981) 34.
30. E. Meinhart and P. Schreier, *Milchwissenschaft* **41**(11) (1986) 689.
31. T.H. Parliment, M.G. Kolor and D.J. Rizzo, *J. Agric. Food Chem.* **30** (1982) 1006.
32. J.P. Dumont, C. Degas and J. Adda, *Le Lait* **56** (1976) 177.
33. L. Siapantas, in *The Quality of Foods and Beverages—Proc. 2nd Int. Flavour Conf., Athens*, ed. G. Charalambous, Academic Press, New York (1981) p. 327.
34. G.E. Mitchell, *Aust. J. Dairy Technol.* **36**(1) (1981) 21.
35. R. Liardon, J.O. Bosset and B. Blanc, *Lebensm. -Wiss. Technol.* **15** (1982) 143.
36. J.O. Bosset and B. Blanc, *Lebensm. -Wiss. Technol.* **17** (1984) 359.
37. K.H. Ney, *Fette, Seifen, Anstrichm.* **87** (1985) 289.
38. E.L. Rondenet and J.V. Ziemba, *Food Engineering*, October (1970).
39. *Food and Drink Manufacture—Good Manufacturing Practice: A Guide to its Responsible Management*, 2nd edn., IFST (UK), London (1989).

Chapter 10

1. *Official Journal of the European Communities.* L. 184 31 (1988) 62.
2. R.A. Boyer, U.S. Patent 2,682,466 (1954); R.A. Boyer, British Patent 699,692 (1953).
3. E.C. Crocker, *Food Res.* **13** (1948) 179.
4. C.G. May, *Food Trade Rev.* **44** (1974) 7.
5. I.D. Morton, P. Akroyd and C.G. May, British Patent 836,694 (1960); C.G. May and P. Akroyd, British Patent 855,350 (1960); C.G. May, British Patent 858,333 (1961); C.G. May and I.D. Morton, British Patent 858,660 (1961).
6. L.C. Maillard, *Compt. Rend.* **154** (1912) 66.
7. C. Eriksson, ed., in *Progress in Food and Nutrition Science*, Vol. 5, Pergamon, Oxford (1981) p. 1; G.R. Waller and M.S. Feather eds., in *ACS Symp.* 215, ACS Washington (1983); M. Fujimaki, M. Namiki and H. Kato, eds., in *Developments in Food Science*, Vol. 13, Elsevier, Amesterdam (1986).
8. T.M. Reynolds, *Food Technol. Aust.* **22** (1970) 610; R. Tressl and R. Renner, *Dtsch. Lebensm. Rundsch.* **72** (1976) 37; R. Tressl, *Monatsschr. Brau* **32** (1979) 240; W. Baltes, *Lebensmittelchemie Gerichtl. Chemie* **34** (1980) 39; R.F. Hurrell, in *Food Flavours*, part A, eds. I.D. Morton and A.J. MacLeod, *Developments in Food Science*, Vol. 3A, Elsevier, Amsterdam (1982) p. 399; L.L. Buckholz, *Cereal Foods World* **33** (1988) 547; H.E. Nursten, *Food Chem.* **6** (1981) 263; H.E. Nursten, in *Developments in Food Flavours*, eds. G.G. Birch and M.G. Lindley, Elsevier, London (1987) p. 173.
9. T.F. Stewart, *Scientific and Technical Survey* No. 61, Leatherhead Food R.A. (1969).
10. J.E. Hodge, *J. Agric. Food Chem.* **1** (1953) 928; J.E. Hodge, in *Chemistry and Physiology of Flavors*, eds. H.W. Schultz, E.A. Day and L.M. Libbey, AVI, Westport, CT (1967) p. 465.
11. Proceedings of Seventh World Congress of Science and Technology, in *Trends in Food Science*, part 1, eds. W.S. Lien and C.W. Foo, Singapore Inst. Food Sci. Tech. (1989) p. 3.
12. J.M. Ames, *Chem. Ind.* (1988) 558.
13. C.H.T. Tonsbeek, A.J. Plancken and T. v.d. Weerdhof, *J. Agric. Food Chem.* **16** (1968) 1016.
14. S.S. Chang, C. Hirai, B.R. Reddy, K.O. Hertz, A. Kato and G. Simpa, *Chem. Ind.* (1968) 1639.
15. H.W. Brinkman, H. Copier, J.J.M. de Leuw and S.B. Tjan. *J. Agric. Food Chem.* **20** (1972) 177.
16. G. MacLeod, M. Seyyedain-Ardebili, *CRC Crit. Rev. Food Sci. Nutr.* **14** (4) (1981) 309; K.O. Herz and S.S. Chang, *Adv. Food Res.* **18** (1970) 1; B.K. Dwivedi, *CRC Crit. Rev. Food Technol.* **5** (1975) 487; G. Ohloff and I. Flament, *Prog. Chem. Org. Nat. Prod.* **36** (1978) 231.
17. G. MacLeod, in *Developments in Food Flavours*, eds. G.G. Birch and M.G. Lindley, Elsevier, London (1986).

18. A. Hobson-Frohock, *J. Sci. Food. Agric.* **21** (1970) 152; A.E. Wasserman, *J. Agric. Food Chem.* **20** (1972) 737; R.A. Wilson and I. Katz, *J. Agric. Food Chem.* **20** (1972) 741; C.G. Janney, K.K. Hale and H.C. Higman, *Poultry Sci.* **53** (1974) 1758; R.J. Horvat, *J. Agric. Food Chem.* **24** (1976) 953; H.S. Ramaswamy and J.F. Richards, *Can. Inst. Food Sci. Technol. J.* **15** (1982) 7; J. Tang. Q.Z. Jin, G.-H. Shen. C.-T. Ho and S.S. Chang, *J. Agric. Food Chem.* **31** (1983) 1287.

19. V.M. Gorbatov and Y.N. Lyaskovskaya, *Meat Sci.* **4** (1980) 209.

20. C.-T. Ho, K.N. Lee and Q.Z. Jin, *J. Agric. Food Chem.* **31** (1983) 336; C.K. Shu, B.D. Mookherjee and M.H. Vock, U.S. Patent 4,234,616 (1980); C.K. Shu, B.D. Mookherjee, H.A. Bondarovich and M.L. Hagedorn, *J. Agric. Food Chem.* **33** (1985) 130.

21. R.J. Park, K.F. Murray and G. Stanley, *Chem. Ind.* (1974) 380; R.G. Buttery, L.C. Ling, R. Teranishi and T.R. Mon, *J. Agric. Food Chem.* **25** (1977) 1227; D.A. Cramer, *Food Technol.* **37** (1983) 249; L.N. Nixon, E. Wong, C.B. Johnson and E.J. Birch, *J. Agric. Food Chem.* **27** (1979) 355.

22. Y. Zhang, M. Chien and C.-T. Ho, *J. Agric. Food Chem.* **36** (1988) 992.

23. J.L. Godman and D.R.D. Osborne, British Patent 1,285,568 (1972).

24. G.A.M. v.d. Ouweland and H.G. Peer, British Patent 1,283,913 (1972).

25. G.A.M. v.d. Ouweland and H.G. Peer, *J. Agric. Food Chem.* **23** (1975) 501.

26. T. Wieland and H. Wiegandt, *Angew. Chem.* **67** (1955) 399; H. Brockmann and B. Franck, *Angew. Chem.* **67** (1955) 303; A. Niewiarowicz and H. Weislo, *Przemysl Spozywizy* **11** (1957) 217; (*Chem. Zentralblatt.* **129** (1958) 8777).

27. H. Sulser, J. DePizzol and W. Büchi, *J. Food Sci.* **32** (1967) 611; See also C.H. Manley and I.S. Fagerson, *J. Food Sci.* **35** (1970) 286; C.H. Manley and I.S. Fagerson, *J. Agric. Food Chem.* **18** (1970) 340.

28. K. Prendergast, *Food Trade Rev.* **44** (1974) 14.

29. C.G. May and C.J. Soeters, British Patent 1,130,631 (1968).

30. R.A. Lawrie, *Food Flavour Ingred. Proc. Pack.* **4** (1982) 11.

31. P.D. Harkes and W.J. Begemann, *J. Am. Oil Chem. Soc.* **51** (1974) 356.

32. W.J. Begemann and P.D. Harkes, U.S. Patent 3,821,421 (1974).

33. D.A. VanDorp, P. Akroyd and L. Mindt, British Patent 1,034,352 (1966).

34. C.J. Mussinan, C. Giacino and J.P. Walradt, U.S. Patent 3,876,809 (1975).

35. I. Katz, R.A. Wilson and C. Giacino, U.S. Patent 3,681,088 (1972).

36. R.A. Wilson, M.H. Vock, I. Katz and E.J. Shuster, British Patent 1,364,747 (1974).

37. P.A.T. Swoboda, AGFD 80,160th ACS Meeting, Chicago (1970)

38. I. Noleau and B. Toulemonde, *Lebensm. -Wiss. Technol.* **21** (1988) 195.

39. G. Giacino U.S. Patent 3,394,015 (1968); U.S. Patent 3,519,437 (1970).

40. C.K. Shu, B.D. Mookherjee, H.A. Bondarovich and M.L. Hagedorn, *J. Agric. Food Chem.* **33** (1985) 130.

41. J.R. Feldman and J.H. Berg, U.S. Patent 3,803,330 (1974).

42. E. Wong, L.N. Nixon and C.B. Johnson, *J. Agric. Food Chem.* **23** (1975) 495.

43. L.N. Nixon, E. Wong, C.B. Johnson and E.J. Birch, *J. Agric. Food Chem.* **27** (1979) 355.

44. R.G. Buttery, L.C. Ling, R. Teranishi and T.R. Mon, *J. Agric. Food Chem.* **25** (1977) 1227.

45. F. Caporaso, J.D. Sink, P.S. Dimick, C.J. Mussinan and A. Sanderson, *J. Agric. Food Chem.* **25** (1977) 1230; D.A. Cramer, *Food Technol.* **37** (1983) 249.

46. H.B. Heath, in *Source Book of Flavors*, AVI, Westport, CT (1981).

47. E. Guenther, in *The Essential Oils*, Vols. I–VI, D. Van Nostrand, New York (1948–1952).

48. J. Velíšek, J. Davídek, J. Hajšlová, V. Kubelka, G. Janíček and B. Mánková, *Z. Lebensm. -Unters Forsch.* **167** (1978) 241; J. Velíšek, J. Davídek, V. Kubelka, J. Bartošová. A. Tučková, J. Hajšlová and G. Janíček, *Lebensm -Wiss. Technol.* **12** (1979) 234; J. Velisek, J. Davídek, V. Kubelka, G. Janíček, Z. Svobodova and Z. Simicová, *J. Agric. Food Chem.* **28** (1980) 1142; J. Davídek, J. Velíšek, V. Kubelka, G. Janíček and Z. Šimicová, *Z. Lebensm. -Unters Forsch.* **171** (1980) 14.

49. Federal Health Authority (Bundesgesundheitsamt) of W. Germany 1984, Bestimmung von 1,3-dichloropropanol u. Verband der Suppenindustrie.

50. D. Chase, *Food Flavour Ingred. Proc. Pack.* **11** (1989) 67.

51. *Evaluation of the Health Aspects of Protein Hydrolysates as Food Ingredients*, SCOGS, Life Sciences Research Office, Fed. Amer. Soc. for Exptl. Biol., Maryland (1978).

52. A.J. Miller, *Food Technol.* **39** (1985) 75.
53. M. Scheutwinkel-Reich, G. Reese and W. v.d. Hude, *Z. Lebensm. -Unters Forsch.* **180** (1985) 207.
54. W.D. Powri, C.H. Wu and H.F. Stich, in *Carcinogens and Mutations in the Environment*, ed. H.F. Stich, CRC Press, Florida (1983) p. 121.
55. Council of Europe, 2nd Internat. Consultation on Flavours, Strasbourg, 27–28th April, 1989.

Index

Printed in Great Britain
by Amazon

35962888R00183